AF575647

AVIATION

C-141 Starlifter

Lockheed's Cold War Strategic Airlifter

JOHN GOURLEY

SCHIFFER MILITARY
4880 Lower Valley Road Atglen, PA 19310

Library of Congress Control Number: 2020943643

Designed by Justin Watkinson
Type set in Impact/Minion Pro/Univers LT Std

ISBN: 978-0-7643-6172-2
Printed in China

Published by Schiffer Publishing, Ltd.
4880 Lower Valley Road
Atglen, PA 19310
Phone: (610) 593-1777; Fax: (610) 593-2002
E-mail: Info@schifferbooks.com
www.schifferbooks.com

Dedication

To all of those in the C-141 community who created, designed, manufactured components, rode in, jumped from, kept in flying condition, improved on with a new variant, flew, and had their lives changed for the better, up to and including survivors of natural disasters and former prisoners of war, as well as those who keep the spirit of the Starlifter alive today, this book is for you.

Acknowledgments

The author was able to, with the generous and extensive help of the following individuals along with some worthwhile internet sites, create this new updated work with unique photographs, line drawings, and data. All of this has resulted in what I hope will be a worthwhile and handy reference source on the Starlifter, as well as providing a book full of interesting and revealing images.

My thanks to Mark Aldrich, Lou Drendel, Dennis Jenkins, Paul Minert, Norm Taylor, and John Vadas. Mike Novack of the C-141 Heaven website provided me with data such as the C-141 timeline and the very useful C-141 Handbook. Other notable web locations include the San Diego Air and Space Museum (SDASM), the Digital Public Library of America (DPLA), and the Defense Visual Information Distribution Service (DVIDS).

Contents

Introduction

The C-141 Starlifter: built in the mid-1960s, there are none flying in regular USAF service at present. One big reason for this was that the fleet was simply worn out, used to the limits of its capabilities each and every day. Missions were as varied as Antarctic resupply flights; humanitarian-aid relief; tactical airlift of cargoes such as vehicles, helicopters, artillery pieces with crews and ammunition, palletized loads of everything from food to tents and jeeps; paratroop drops; low-level special-operations missions; airborne command post functions; disaster relief; transporting missiles, rockets, astronauts, and their spacecraft; as well as returning former prisoners of war back home. C-141s were also highly adaptable, versatile platforms. It replaced many types that were either too slow or in need of retirement. It introduced jet-powered propulsion in podded engines attached underneath the wings, an unobstructed cargo cabin, pressurization for high-altitude flying, and a cargo ramp that could be opened in flight to air-drop materials without landing to unload them.

After its introduction to use with the Military Air Transport Service (MATS), later to be renamed Military Airlift Command (MAC), it was determined in many cases that the C-141A was capable of greater cargo lifting, but due to its existing cargo cabin length, maximum loads could not be carried. To address this shortcoming, a program was undertaken in the 1970s to lengthen the fuselage, while adding the capability to refuel while in flight. This C-141B variant made full use of the hauling abilities the plane was truly capable of, while achieving increased global airlift endurance limits with the midair-refueling package installed.

C-141s not only were put to use here in the continental US but provided their services in places such as Southeast Asia, Panama, the Middle East, Africa, and the South Pole and were a familiar sight at air bases and airports the world over, supporting exercises, responding to calls for aid, and being used in aeromedical evacuation flights. The aircraft would have played a crucial role in the event that NATO needed rapid material reinforcement in the wake of a Warsaw Pact invasion of western Europe during the Cold War era. It proved its ability in actual battle conditions in the 1973 Yom Kippur War, as an air bridge was established, flying precision routes into and out of the state of Israel. It did so again in real-world events such as Operations Desert Shield/Storm and was even used as platforms for exotic items such as high-powered optical telescopes and in early research efforts into infrared-countermeasures development.

Other versions included the C-141B SOLL II (special operations, low level), used for precision airdrops at various speeds, hugging the terrain to achieve the element of surprise and avoid detection while dropping equipment vital to special-ops teams to accomplish their objectives. The C-141C was a further enhancement of the C-141B model, introducing the EFIS (electronic flight instrumentation system or "glass" cockpit) modern air traffic control identification and terrain avoidance gear.

Today, with the C-17A Globemaster III in large numbers, filling the needs of our modern-day Air Mobility Command (AMC), the Starlifter has been largely forgotten. But to those who built and flew it, maintained it, jumped out of it, rigged and released cargo of many configurations and varieties while in midair, and brought victims of famine and storm devastation much-needed and much-appreciated relief and comfort while evacuating those who survived such events, as well as those who were liberated from the horrors of forced captivity, the C-141 will forever be remembered as one of the greatest airlifters ever conceived. Versatile, adapted to many mission needs, and modified to give even more during its midlife existence, the C-141 was one of those designs that gave the American taxpayers everything that was expected of it, and then some. Despite most of the fleet being scrapped for salvage purposes, several survive today in museums and in air parks across America, none more famous than the "Hanoi Taxi," 66-0177, proudly preserved and displayed at the National Museum of the United States Air Force in Dayton, Ohio.

CHAPTER 1

C-141A Background History

As the United States Air Force entered the 1960s, it had many aviation assets dedicated to the roles of airlifting military-related cargo. Planes such as the C-74 Globemaster I, C-124 Globemaster II, KC-97 Stratofreighter, C-130 Hercules, and C-133 Cargomaster all had functions, from hauling paratroops to air-dropping cargo to carrying outsized loads. With the introduction of jet propulsion, faster speeds and greater range were attainable. Large airframes, such as the B-47 Stratojet and B-52 Stratofortress, proved that jet engines were the power plant of choice for a new transport plane. This was further reinforced by service entry of the C-135 Stratolifter, but it was effective only up to a point. Its low wing design, with a side-loading door high on the left side of the fuselage, was not optimized for loading or removing heavy, cumbersome freight. An all-new aircraft design dedicated to the transport role, to supplement and eventually replace the propeller-driven types, was needed.

By the late 1950s, talk was of a new plane that could fit the requirements not only of strategic airlift needs but also one of supporting situations involving limited conflict, so-called "brushfire wars." The United States Army pushed this idea, stating that new airlifters should possess the capability to quickly transport heavily armed, mobile combat forces fast enough to contain and eliminate an opposing armed force while aggression was still in its incipient stages of execution. This idea would later coalesce into the "flexible response" concept. Operations ranging from flights over the Himalayas (or "the Hump") to support forces in China during the Second World War to Korean Conflict experience showed that a dedicated cargo hauler could add options in airlift not available previously.

To justify development of a new transport, it also needed to be compatible with commercial operators, with the means to be modified from civil use to military needs as situations called for. In 1952, discussions began in earnest for an all-jet transport, one capable of intertheater (moving from one region of the globe to another) as well as intratheater (positioning military assets within the contested region itself) movement of cargo or troops. With conversations ongoing, emphasis leaned toward more of a civil requirements priority, with military considerations taking a back-burner status. This resulted in support for the Military Air Transport Service (MATS) eroding, since the reasoning went that a new fleet of civil-oriented airlifters could quickly be modified and pressed into military service if a conflict erupted. This concept gained traction in Congress, where members proposed offering government loans to civil airlines as a way for them to build up enough of these planes so that augmenting-crisis needs could be met if required. MATS would then consist of a small fleet of planes, enough to perform taskings short of conflict.

By 1959, MATS was under threat, with no definitive strategy for airlift on the horizon. Cmdr. William H. Tunner of MATS proposed the following items for a new airlifter that got things back in motion in a logical sense. He put forth that a big plane was needed, one that could move outsized cargo that did not need to be disassembled or cut into pieces and then rebuilt in the theater of action, as had been done in past military campaigns. Cargo had to be loaded "straight in" through open access to the fuselage from the front or rear. Cargoes including the biggest missile we had in the inventory should fit inside. The cargo floor itself had to be the height of the loading trucks delivering items to be flown out. On the basis of the above, the plane had to have a wing mounted high on the fuselage section. Range had to be on the order of 7,000 miles. This would offer maximum options for loads heading west over the Pacific from California to places such as Hawaii and on to Japan, for example. Flight speeds were not emphasized, since by using this method of moving material, time was being saved compared to attempting to move everything by surface means. When hoping to move an 80,000-pound payload, piston engines were not to be used. Turboprop or jet power plants were to be installed. By moving loads to and from zones of conflict, costs overall should be reduced to an economical level.

Tunner ended his list by adding, "This is our major requirement. A work-horse plane, cheap to operate, of indifferent speed, relatively large and easy to load. It would be the backbone of the fleet." This list of general items, with the Army needs as well as a congressional committee's findings, led to MATS offering up a QOR (qualitative operational requirement) in February 1959 for a Logistic Aircraft Support System, which was submitted to Headquarters USAF. Of then-currently fielded aircraft capable of filling the role envisaged for the future, only the Douglas C-133 was able to do so until the 1970s. But to realize the true benefits of the new requirement, only a turbojet-powered type could truly fit the need. The QOR went on to spell out some background, such as a plane of medium size, ensuring reasonable frequency of service, capable not only of local airlift missions within the zone of interior, but also having the capability for transoceanic operations. Regarding aircraft designs, the QOR went on to state that then-current designs of large, high-speed, long-range, intercontinental aircraft would not on their own afford the necessary flexibility of operation. This equipment must be able to handle general war requirements. The worldwide short-field and terrain difficulties would seriously limit its value in limited war. For example, studies concerning the usability of the VC-137A indicate that this aircraft can operate from only about 30 percent of the worldwide airfields now used by the 1254th Air Transport Group at Washington National Airport.

Headquarters USAF welcomed the recommendations, but Maj. Gen. James Ferguson, HQ USAF director of requirements, warned that then-current funding levels would not allow for other than a small inventory of large transports to be purchased. Things might be more favorable if the size was kept within the medium-transport dimensions. He did not even foresee a mix of medium and large planes being bought, either, but he did suggest a joint civil-military venture to develop such planes. Short of this, he did state that the Air Force would likely be better off attempting to create a transport by using its available resources. He ended by asking MATS to conduct a study of the benefits of procuring either a medium- or large-sized transport system.

Gen. Tunner replied in March 1959 that "Deletion or omission of either the proposed heavy or medium transport will sooner or later compromise the airlift logistic system." He did agree with the previous comments regarding the C-133, but that immediate development of a "Medium 'work-horse' transport" should be initiated. By June, HQ USAF wrote up a general operational requirement (GOR) for a medium transport and had it further refined by December into a definitive specific operational requirement (SOR), forwarded to MATS. Gen. Tunner was briefed that the SOR met or exceeded MATS requirements as set forth in the QOR dated February 16, 1959. Meanwhile, Gen. Thomas D. White, Air Force chief of staff, and Army chief of staff Gen. Lyman L. Lemnitzer jointly discussed and refined the needs of the Army on the topic of air transport requirements. Opposition to the idea of a new transport plane was fading, and the military exercise "Big Slam / Puerto Pine" revealed shortcomings and equipment obsolescence in the methods used to transport hardware by air.

A congressional hearing by the Rivers Committee, chaired by L. Mendel Rivers from March 8 to April 22, 1960, took an unbiased, objective, and logical look at the subject of airlift as a part of national defense needs, reviewing the "Big Slam / Puerto Pine" critiques. The Air Force supported the SOR outline (soon to be titled SOR-182), with Maj. Gen. Bruce K. Holloway, Air Force director of operational requirements, stating the following to the committee: "Considering existing and expected funding limitations, it is unlikely that more than one cargo aircraft development of this magnitude could be concurrently funded. In recognition of this factor, every effort has been made to evolve aircraft specifications that would meet the military requirements and, additionally, assure [*sic*] an aircraft possessing a high degree of commercial compatibility. Secondly, considering the recognized obsolescence within both the military and civil airlift fleets, the factor of early availability was given considerable weight. As a result, our specific operational requirement stipulates a state-of-the-art aircraft. Lastly, and most important, is the effective capability of the aircraft to be developed. The design of the aircraft must encompass performance and reliability features which will not only assure [*sic*] its effective utilization under widely varying conditions but also assure [*sic*] the continued effectiveness of this aircraft in order to preclude early obsolescence." Gen. Holloway went on to state three paths the Air Force could take: (1) off-the-shelf procurement of an existing cargo aircraft or cargo-modified versions of a jet transport, (2) procurement of a future turboprop-powered plane accepting high development costs, and (3) the favored option: the development of an optimized cargo aircraft.

To this end, Gen. Elwood R. Quesada, the Federal Aviation Administration (FAA) administrator, offered his support for the optimum-aircraft option, dropping his former opposition to buying CL-44 aircraft. He informed the committee that "I want to make it crystal clear that I am a strong supporter of the concept that the Military Air Transport Service should be a modern, basic

The Convair submission to the SOR-182 specification was this plane with engines on a pair of pylons, conventional tail fin, high wing, and large fairings for the main landing gears. *SDASM Archives*

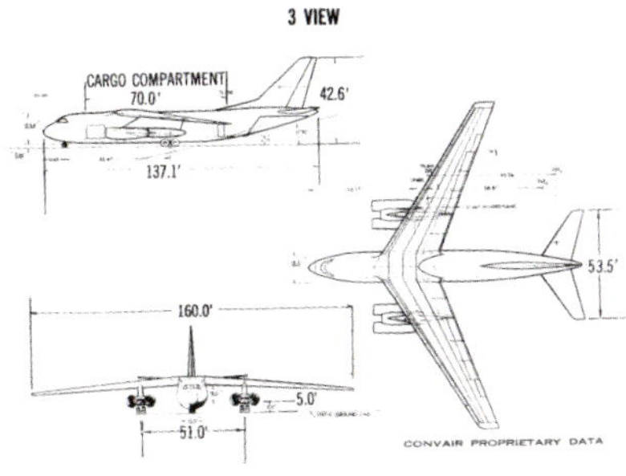

Three-view line drawing showing left-side profile and head-on and topside angles. Cargo loading was to be accomplished by means of access to cabin at the rear of the fuselage. *SDASM Archives*

Underside view shows the main gears, each located away from the fuselage. In doing so, the cargo space is uncluttered and free for straight-in loading of outsized payloads. *SDASM Archives*

With the pair of side doors open, ramp lowered, and upper aft panel raised above truck configured for airdrop, it can now be moved on its pallet into the aircraft. *SDASM Archives*

weapons system in support of military missions . . . my concept of a highly trained, instantly ready MATS, equipped with truly modern equipment, is confined to a fleet designed expressly and solely for effective performance of a clearly defined, hardcore military mission. If we compromise the military fleet now, it will be compromised for the next decade at least."

He added that instead of buying Boeing 707 or Douglas DC-8 types (or both), he preferred, along with the Army, to obtain Lockheed C-130B Hercules transports instead. The Air Force did not waver, stating that it favored the SOR document outline. Options under consideration for approval swirled around the C-130, the 707, the DC-8, and the SOR-proposed type. As the Rivers Hearings concluded, the recommendation was to approve funding for the SOR development program. This requirement was by now deemed so important that after the inauguration of President John F. Kennedy, his first official act was to initiate the order to draft a document detailing the specifications the new transport had to fulfill, known as Specific Operational Requirement 182 (SOR-182), issued on May 4, 1960. At only eight pages, it (as well as an amended document of four pages, SOR-182-1, released on August 15, 1960) spelled out desired features for a military and civil version; for example, the ability to operate into and out of established military bases throughout the world. The plane must be able to make a 180-degree turn on a runway of 150 feet in width. Taking off and landing while carrying an internal payload of from 50,000 to 60,000 pounds, and being capable of a 4,000-mile flight, should be safely accomplished while operating on a 6,000-foot or shorter runway.

The aircraft would be used in all weather conditions and be capable of inter- as well as intratheater positioning for military purposes, for the transport of cargo in domestic and international routes in the civil variant. Access to components for inspection and ease of maintenance was stressed, and the ability to change out electronic systems was also emphasized. Structural capacity of the aircraft should be approximately 70,000 pounds. The extra crew compartment, cargo ramp, and cabin floor stress values should be at least 200 pounds per square foot, and crushing strength should be 750 pounds per square inch. Cargo cube accommodation should be at least 6,000 cubic feet. Bulky cargo capacity for one large unit of cargo or several smaller ones to be safely loaded and secured required an unobstructed rectangular area of 70 feet long by 10 feet wide and 9 feet high. With such cargo onboard, a safety aisle of at least 14 inches in width must be maintained throughout the entire length of the cargo bay, for personnel to pass freely to check secured cargo and for expeditious, unhindered movement to doors / escape hatch locations in an emergency. A cargo-loading method of straight in through the tail section was needed. Truck bed/trailer loading/unloading height was 48 inches, most efficient overall to expedite cargo movement, with the ability to adjust the cargo floor ramp to ground level within ±1.5 degrees. A loading ramp would lower to facilitate vehicle loading/unloading, with an incline of no more than 11 degrees. A stated need for a side cargo door for secondary loading, measuring 108 inches wide by 78 inches high, to be located at the forward section of the fuselage, was deleted in the amended SOR-182-1 document. Handling gear (when stowed, not to intrude into the cargo space area, of up to 4,000 pounds) would be compatible with standard 463L pallets.

All cargo must be capable of air-dropping or being jettisoned in flight at reduced airspeeds through the aft doors, which were to be capable of being opened and remaining stable/secure when exposed to the airstream. To maximize aircrew utilization on long flights, a relief crew compartment could be installed at the forward end of the cargo bay and removed quickly for short flights with large cargo loads. It had to contain three rest bunks, a galley, trash containment, and flight gear storage within a 7-foot module. In an emergency requiring the use of oxygen, the aircrew (consisting of pilot, copilot, flight engineer, navigator, and a possible fifth flight check member riding in the cockpit jump seat) had to have a supply provided so that they would be able to stay at their assigned stations. Troops/passengers occupying plug-in-style seats attached to the cargo cabin floor could use portable oxygen devices. All should have enough to breathe for five hours' duration.

Three latrines were needed: two in the cargo compartment and one on the flight deck (later moved to the forward end of the cargo bay). Litter racks for up to forty-two patients should occupy the center section of the cargo space, of two-row and triple-tier arrangement. This configuration shall be of minimal effort to set up and not compromise the cargo space cross-section design. Cabin soundproofing/heating/ventilation should meet USAF/FAA conditions. Military C-141s would have extra sound insulation to allow occupants not to have to wear hearing protection, up to a ten-hour flight maximum. Heating interior spaces must be maintained at austere locations, to prevent sensitive cargo from freezing and to provide a steady source of warmth for aircrews and passengers. Enough fuel for a 4,000-mile flight (3,000 miles for civil example) with reserves had to be contained within the wings. Extra fuel for trans-Pacific hauls, adding 1,500 miles more range, could be provided for but not stored in the cargo section or adjacent to the aircrew area. Cargo displacement could be reduced to increase range, but the reduction could not exceed a load factor of 2.25. Cruise altitude was at least 25,000 feet (with cabin pressure maintained to the equivalent of 8,000 feet), mainly in order to avoid weather conditions at lower levels. The design had to take advantage of the best possible transit speeds in the cruise phase. While in traffic pattern routes near airfields, speeds were acceptable in accordance with aircraft then in use. Communications were needed for worldwide contact with command posts, as well as for normal uses in peacetime operations. Such equipment would include one VHF radio (ARC-73 type), two single sideband, and three UHF. Internal contact was via four interphone/public-address devices used by the pilot, loadmaster, jumpmaster, cabin passengers, and exterior ground crew. Navigation gear, suited to FAA and Air Force models, would be installed. Military outfit would include dual low-frequency ADF, dual VOR/ILS, dual TACAN, dual Doppler nav system, weather contour radar, and dual ATC transponders. As the airframes were being constructed, antennas, connectors, and wiring were to be installed. In an effort to meet FAA civil guidelines, the following items were stipulated: cockpit visibility was to meet FAA regulations, additional thrust methods other than water injection were not desired, thrust reversers were to be included, and engine sound suppression, fuel dumping, and single-point pressure refueling were needed. If conflicts in meeting the military/civil dual track were not possible, the military needs would take precedence. A flight simulator and instrument procedures training capability unique to the C-141 were mandated.

Congress approved Public Law 86-601 in the amount of $310,788,000 in July 1960 for the Air Force to buy or modify aircraft it currently operated, but $50,000,000 had to be spent on development of an aircraft as specified in SOR-182. Proposals were sent out in December to Boeing, which submitted their Model 731 concept, to Convair, to Douglas, with their Model 2085, and to Lockheed, these facilities being judged the best at eventually producing an acceptable product. The four contractors in January presented oral briefs to the Air Force regarding the new aircraft. The winner, announced by President Kennedy in March 1961, was the Lockheed Corporation, their Model GL-207-45 Super Hercules being the best design to fulfill the stipulations laid out in SOR-182. The USAF officially labeled the new plane C-141, previously referred to as System 476L. In April, the engine designation of TF 33-P-7 (known commercially as the JT3D-8A) was approved, and the budget plan for the next six fiscal years (FY61 through FY66), for a total of $1.079 billion for the total program, was authorized. Simultaneously, Air Force contract AF33 (600)-42921 paid $29.9 million for the five RDT&E (research test development and evaluation) airframes to be assembled. The sum of $100,000 was used to purchase needed aerospace ground equipment. In May, the C-141 Special Programs Office (SPO), in agreement with MATS, decided to delete two items in the design.

Taken out of the airframe were aerial-refueling capability and the autoland functions. June saw the addition of three mobile training units (MTUs) and three flight simulators. In July, amendment #1 was introduced to the contract. It stated that a structural-integrity program be carried out that conducted a 100 percent design load limit, a static test program, a complete flight loads survey program, and a fatigue test program based on a life cycle of 3,000 flight hours and 12,000 landings. In August, the model designation changed from GL-207-45 to 300-50A-01. The *A* designation was added to the C-141. Aircraft coming off the production line would be labeled C-141A-LM (the LM signifying built at Lockheed Marietta), with three of the five planes now set for USAF evaluations. Long-range plans included notifying the Eglin Air Force Base (AFB) McKinley Climatic Laboratory of the plane arriving in September through November 1964. Engine cold-weather tests there would be conducted in November–December 1963. In September, the USAF announced subcontractors for the assembly of major components. These included Bendix, which would supply the main landing gear, and Cleveland Pneumatic, which would provide the nose landing gear. The empennage fabrication was to be accomplished by General Dynamics. In November, due to concerns about weight increases, Lockheed proposed forty-three changes to be made in the assembly process. In December, the Langley wind tunnel flew a scale model to investigate aerodynamic flutter. In January 1962, the C-141 mockup was inspected, with 127 requests for approval reviewed. Approval to act on ninety-eight of the items was agreed upon. Further announcements of subcontractors providing components to the project included Avco, which would supply the wing box beam, Eclipse-Pioneer, which would deliver the automatic flight control system (AFCS), AiResearch, which was scheduled to manufacture the environmental control system, Goodyear, which would supply the brakes, and Bell Aerosystems, which would produce the cargo compartment floor plates. February saw a reversal of an earlier decision, in that an automatic landing system would now be incorporated into the avionics suite. Another mockup inspection was also done, this time by the FAA, with eleven US and foreign airline representatives included. March saw a response from seventy-eight bidders to the request for proposals (RFP) to build the flight simulators. In April, Twin Industries was awarded the contract to build wing trailing edges. An Air Force letter contract, AF 33(657)-8835, was issued in May 1962 for the purchase of sixteen more aircraft. June saw a study conducted into the C-141 having the ability to carry the SM-80 (Minuteman) intercontinental ballistic missile (ICBM). Curtiss-Wright was awarded Air Force contract AF 33(657)-9523 to build the C-141 flight simulators, with two to be bought in FY63. Lockheed also was presented with letter contract AF 33 (657)-8957 for the mobile training units. July saw completion of the first major subassembly, the nose landing gear and housing, as well as delivery of the first XTF 33-P-7 engine to Lockheed. The C-141A technical manuals were completed in August. Lockheed ran the XTF 33-P-7 engine for twenty-eight hours, twenty-seven minutes. Plans to further study C-141A/Minuteman transport feasibility were approved for more investigation in September. Lockheed ran the first XTF 33 engine for up to forty-two hours in October as the second engine was delivered. Engine climatic tests at Eglin began in November. Assembly of fuselage sections began in December, with engine climatic testing revealing the need to add a fuel enrichment device for cold-weather starting. The aft, mid-, and forward fuselage sections were joined together in January 1963, and an engine endurance test of 150 hours was completed. In February, Avco delivered the first wing box beam. The number 1 fuselage was pressure-tested in March, and the first production engines were delivered to Rohr for checkout and installing the engine cowlings and thrust reversers. April saw installation of the wings, landing gear, and cargo ramp assemblies to airframe #1, with wing trailing edges and associated subsystems installed in May, with a fixed-price contract of $6,989,613, AF 33(657)-10383, awarded to the Link Division for six flight simulators. In June, the German air force and Lufthansa contacted Lockheed about purchasing a single C-141 in FY65. In July, a Lockheed cost proposal for 127 aircraft was forwarded to the Air Force. On August 22, President Kennedy made an announcement by reading a prepared statement on the introduction of the C-141, then pushed a button on his Oval Office desk, allowing the hangar doors to open at the Lockheed plant. With this, 61-2775 was rolled out one week ahead of schedule to mark the ceremonious introduction of the plane. Five days later, the Air Force accepted the first Starlifter. In October, the USAF accepted the second plane, with the first flight of 61-2775 taking place in December, consisting of fifty-five minutes in the air.

A follow-on contract in May 1964 to Lockheed was for 122 more C-141As. In June, the first plane was flown to Edwards AFB for more-comprehensive testing.

Comparison of the C-141 to the C-130 Hercules shows the longer fuselage, swept wings with jet engines, and T-tail layout. Both could essentially perform the same missions, but the Starlifter could do it faster, could carry more, and could also contain within its cargo bay one Minuteman ICBM sealed in a transport container. A C-141 once air-dropped a cargo load of 70,195 pounds, a world record. *Lockheed*

11:00 a.m.
August 22, 1963

TCS DRAFT
8/16/63

MARIETTA, GEORGIA CEREMONY

I am delighted to have the privilege of pressing the button which will roll-out the first all-jet transport C-141A to come off the assembly line at Marietta, Georgia. I am delighted, first of all, to be associated in this act with two distinguished sons of Georgia whose careers have been devoted to strengthening America's security -- Senator Richard Russell and Congressman Carl Vinson.

I am delighted, secondly, because this symbolic act reminds us of many fundamental truths. Marietta symbolizes the role played by so many communities, large and small, North and South, in contributing to the military might of the United States. ~~The plant which produced these~~ planes, along with its facilities and its payroll, also symbolizes ~~the role played by~~ the United States in ~~the economic life of so many~~ communities.

And this first C-141A symbolizes not only the modernization of our airlift fleet but the overall strengthening of this Nation's defense posture which has been underway for over two years. We have greatly increased our nuclear retaliatory power, the number of protected missiles and alert aircraft that can strike so devastating a blow in response to aggression ~~an attack~~ that no power on earth could rationally choose to attack us. We have greatly increased our conventional forces, their firepower, their versatility, their equipment, and our ability to transport them quickly to any spot on earth where peace and freedom are endangered.

Above: The prepared statement read by President John F. Kennedy on the day the C-141 Starlifter was rolled out. He realized the potential of and took a keen interest in this modern, jet-powered addition to the nation's cargo-carrying capacity. *To right*, a painted scene of netting being secured over wooden crates as a C-141 takes off on an airlift mission. The angle shows off the long, wide fuselage; high tail fin; engine nacelles; and large fairings that contain the landing gear.
Above image, JFK Presidential Library; right image, Lockheed

Two artist's renderings of the new C-141. Both are very much like the final C-141A as built. The large cargo bay, high wing, engines separated from the fuselage, stabilizers mounted on top of the fin, and easy ground access to load/unload all were revolutionary features as compared to older cargo/transport aircraft in the inventory. Black stripe with MATS in white stands for Military Air Transport Service. Dayglo orange around nose and aft fuselage aid in easy visual sighting when used in regions such as the Arctic or Antarctic. *Both images, Lockheed*

Rollout ceremony for the C-141A Starlifter at the Lockheed Georgia plant in Marietta, on August 22, 1963. The name Starlifter resulted from a Lockheed employees' in-house name selection contest. The troops and jeeps on the ramp were just a small sample of what the plane could carry internally. *Lockheed*

The Starlifter took to the air on its maiden flight on December 17, 1963, the sixtieth anniversary of the Wright brothers' first flight. The all-Lockheed flight crew was, *from left to right*, A. P. "Bob" Brennan, flight engineer; E. Mittendorf, flight test engineer; Leo Sullivan, chief engineering test pilot, and pilot on this first flight; and Henry "Hank" Dees, copilot. The successful milestone of just under an hour, reaching an altitude of 7,800 feet, originating and ending at Dobbins AFB, Georgia. *Lockheed*

C-141A, 61-2775, in straight, low-level flight profile while photographed from a chase aircraft flying above. High placement of the wing was one design feature allowing for a large, uninterrupted cargo cabin section. Trailing-edge devices, the flaps and ailerons, are unpainted in this picture. Nose probe contains flight instruments used to cross-check and verify readings from the installed aircraft systems. Small fins seen at wing and stabilizer tips were used to measure airflow patterns at these locations. *Lockheed*

The second C-141A, 61-2776, flew to Edwards AFB for more-comprehensive testing. Fuselage is unpainted. MATS emblem is aft of star and bar roundel. MATS is in white letters on a black background stripe across tail fin. Black checkerboard squares are photocalibration markings, used to provide known reference points for future study of photographs taken as tests proceed. Calibration marking on tail fin replaces the American flag. Bureau Numbers (BuNos) are five digits in black. The manufacturer's serial number, 6002, is on the nose. *USAF*

In the left picture, C-141A MSN 6002, 61-2776, has an instrumentation probe mounted on the radome. The sensors measure dynamic factors such as airspeed, air density, angle of attack, and wind gusts. In the image above, the earliest Starlifters were built with upward-hinging normal crew access doors. In an emergency, the door was jettisoned and the crew departed the plane by means of an escape chute. Later on, this door would be altered to one that moved into and up inside the cabin. *Both photos, Paul Minert collection*

C-141A, MSN 6112 / BuNo 65-0260, being assembled at the Lockheed Georgia plant. Radar is yet to be installed. Work platforms surround the engine nacelles. *Lockheed*

61-2775 in final stages of assembly. Nose section to left is used to test-fit and check out hardware before it actually is installed into the production aircraft. *Lockheed*

MSN (manufacturer's serial number) 6001/61-2775 in view as work proceeds. Landing-gear fairings are to right, and wing box is already on top of fuselage. *Lockheed*

With fuselage raised at nose, the tail fin is in the process of being attached. Aft section cutout is where the pair of petal doors will be fitted. *Lockheed*

A C-141A tail fin is completed and ready to be mounted to an aft fuselage section. The upper horizontal surfaces, the stabilizers, are of an all-moving design. They are hinged inside the center of the vertical tail fin, tilting up or down together when control inputs from the cockpit are moved. The spike protruding forward is a high-frequency (HF) radio communications antenna. *Lockheed*

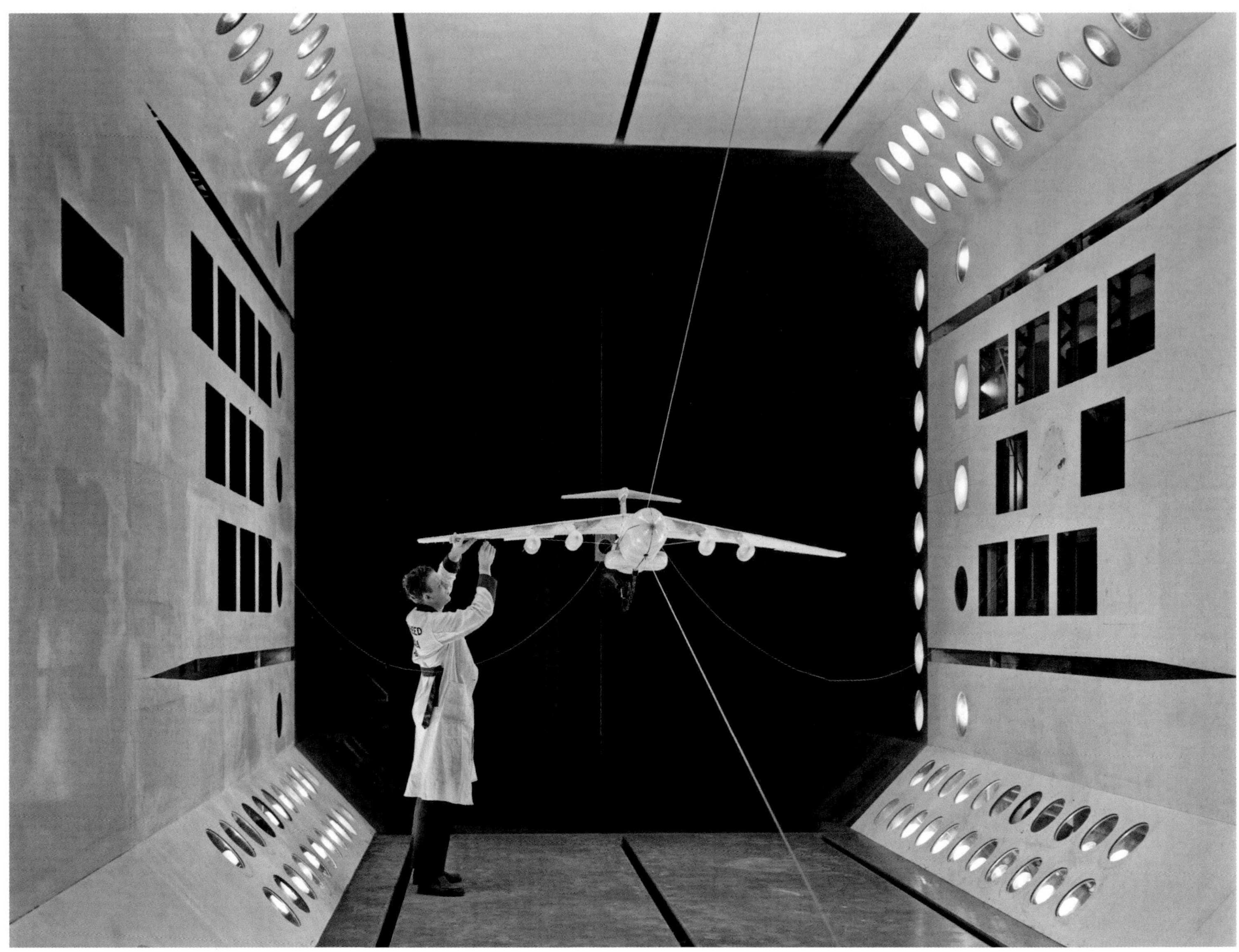

Lockheed conducted extensive wind tunnel testing to verify the C-141 airframe design. While the high wing and T-tail configuration are ideal for straight and level flight, problems can occur at high angles of attack, with loss of directional stability since the turbulent airflow from the wings can affect the input from the tail fin surfaces. In an extreme situation, a deep stall can occur, with the plane becoming uncontrollable. Understanding this phenomenon was a priority in order to develop a safe flight envelope that aircrews must operate in, so this scale model is being "flown" inside a Transonic Dynamics Tunnel (TDT) facility. *National Archives via Dennis Jenkins*

One major advance of the C-141 over all the previous cargo aircraft was the ability to load bulky, outsized cargo through the tail section, with minimal time spent loading/unloading. In the left photo, a mockup aft fuselage was used to perform actual loading operations, refining the process and incorporating changes to be put into practice in the field as events were carried out. At right is the finished product, C-141A, 61-2775, with petal doors opened and ramp lowered as troops wait to board. This design has endured over the years as future airlifters have been built, and has been used with essentially this same configuration. *Both photos, Mark Aldrich collection*

C-141A Starlifter Specifications

Wingspan: 160 feet. Wing is 3,228 square feet, at a sweep of 25 degrees. Fuel tanks contained within wing structure. Upper and lower spoilers can be deployed in flight (upper to 27 degrees, lower to 59 degrees); on landing, spoilers extend farther to 90 degrees upper, 86 degrees lower, to assist in achieving a short landing length. Spoilers can be armed to extend on landing automatically, once landing-gear wheels reach 50 knots' rotation. Can also be linked to throttles, so that if reverse thrust is selected during an aborted takeoff, they will automatically deploy. Flaps are of the double-slotted Fowler type. Flaps and spoilers cover about 60 percent of the wing surfaces.

Length: 145 ft. **Height**: 39.3 ft.

Weights: Maximum ramp: 318,000 lbs. Maximum takeoff weight: 316,600 lbs. Design landing weight: 257,600 lbs. Equipped weight, empty: 136,030 lbs. Design payload: 68,500 lbs.

Performance: Maximum cruise speed: 485 knots. Design cruise speed: 440 knots. Aerial-delivery drop speed: variable, from 115 to 200 knots. Rate of climb at sea level on a standard day, four engines, and normal power at 316,100 lbs.: 3,330 feet per minute. Takeoff distance at 316,600 lbs.: 4,800 feet. Landing distance at 257,500 pounds: 3,500 ft.

Trans-Pacific Ocean flight times (zero wind / international standard atmosphere (ISA), 0.25 hours for ground maneuvers and MATS fuel reserves): McChord AFB to Tachikawa, Japan, 4,171 nautical miles (nm) with 26-ton load: 10.2 hours; McChord to Wake Island, 3,786 nm with a 29.7-ton load: 9.3 hours; Travis AFB to Hickam AFB Hawaii, 2,113 nm with a 34-ton load: 5.3 hours; Hickam to Sydney, Australia, 4,400 nm, with a 23.7-ton load: 10.7 hours; Midway Island to Okinawa, 2,910 nm, with 34-ton load: 7.2 hours; Travis to Wake Island, 3,833 nm, with a 29.5-ton load: 9.4 hours.

Engines: Four Pratt & Whitney TF33-P-7. Each thrust-rated at 21,000 pounds at sea level. Engine length is 142 inches, diameter is 54 inches, and dry weight is 4,490 pounds each. There are sixteen compressor stages, including two fan stages. All engines and nacelles are interchangeable on the aircraft. Access panels provide quick, easy views to components and maintenance accessibility. All engines capable of reverse thrust. Each mounted under the wings, in separate nacelles and attached to wing with aerodynamic pylons.

Fuel system: Type fuel required: JP-4. Capacity: 23,080 gallons / 150,020 lbs. Single-point refueling rate from starboard main landing-gear fairing: 900 US gallons / 5,850 pounds per minute. Fuel distributed throughout ten internal tanks; tanks can also be gravity fed. Fuel jettison rate: 700 gallons / 4,550 pounds per minute. A single manifold controls fueling/defueling, cross-feed, and jettison. Electrically driven fuel boost pumps in each tank feed fuel to the engines.

Oxygen system: One 25-liter liquid oxygen (LOX) converter for the aircrew. Two removable 75-liter LOX converters for the cargo cabin occupants.

Air-conditioning and pressurization: Direct-bleed air from engines to both AC and pressurization systems to maintain a maximum differential air pressure of 8.2 psi.

Hydraulic systems: Three total. Pressure is 3,000 pounds per square inch (psi). Any two operating engines can maintain pressure integrity. System 1 is driven by engines 3 and 4, supplying half of the pressure needed to operate elevator/rudder and aileron actuators. System 2 is driven by engines 1 and 2, delivering the other half of pressure capacity to primary control surfaces, and also powers the landing-gear extend/retract cycles, moves flaps, operates brakes and nosewheel steering, deploys spoilers, horizontal trim tabs, and the 2.5 kVA emergency generator. System 3 is in standby, plus it provides additional power to flaps/spoilers, and main wheel braking backup. It also operates the aft cabin pressure door open/closed, ramp cargo doors, and APU starter.

Electrical system: Primary is 115-volt/400-cycle alternating current (AC) supplied by four 40 kVA alternators, one mounted on each engine. A fifth 40 kVA alternator is part of the auxiliary power unit (APU). APU supplies ground power with engines shut down. Direct current (DC) is supplied by two 200-ampere transformer-rectifiers, converting AC to DC. One 24-volt, 11-ampere-per-hour battery is used mainly for starting up the APU. Plane will operate with two alternators operable if necessary. Emergency power source is provided by the 2.5 kVA AC/DC generator driven by the number 2 hydraulic system.

Deicing system: Electric deicing for leading edge of horizontal stabilizer, and for cockpit window defog/defrost and deice. Engine bleed air for leading edges of wings and engine inlets.

Landing gear: Tricycle-type overall arrangement. Twin nose gear wheels. Nosewheel steering capable of being turned 80 degrees either side of center. Mains of two four-wheel bogies, located externally of the cargo cabin, that free-fall to the down and locked position. Struts compress 17 inches upon touchdown, absorbing energy, then depress 8 inches more for softer landings avoiding aft fuselage striking the runway.
Landing-gear tires: Nose: two 36 × 11 type VII. Tubeless, pressurized to 200 psi. Contact area: 130 square inches. 9.8 inches wide / 16.03 inches in diameter. Main: eight 44 × 16 type VII. Contact area: 225 square inches. Short-haul pressure rating: 150 psi / long-haul pressure rating: 180 psi. Tires are 13.39 inches wide / 20.36 inches in diameter.
Flight-deck/instrumentation: Area of 110 square feet. Contains two bunks, seats for relief crew. Removable flight check / observer seat aft of central control pedestal. Wraparound windshield area is 31 square feet, providing a 135-degree arc of outside view. Engines can be seen from either pilot or copilot seats. Navigator sits behind pilot. Engineer behind copilot. Two forward side windows open and slide aft, for ventilation on the ground and as an emergency means of escaping the aircraft. Instruments consist of vertical tape gauges for altitude, airspeed, and engine functions. A pair of central air data computers (CADC) provide each pilot true indicated airspeed, Mach number, altitude, and vertical speed; automatic flight control system (AFCS) inputs type PB-60, Mach trim compensator, and operation of the ASN-24 navigation computer. Solid-state components include dual radio compasses, VOR/localizer, glide slope, and marker beacon. Standard USAF items include TACAN, Doppler, and weather radar, APN-59, later upgraded to the APS-133. Headings are indicated by dual gyro compasses. LORAN-C gives position fixes, detected by signals from ground stations up to 1,000 miles distant. Push-button ergonomics allow aircrew to switch between navigation modes. Communications systems include identification friend or foe (IFF), internal intercom, public address (PA) system, and dual high-frequency, very high-frequency, and ultra-high-frequency (HF/VHF/UHF) systems.
Personnel/cargo capacity: Flight crew: four. Eight on a long-haul flight with four relief crew. Troop capacity: 154, reduced to 131 if a relief pallet is included. Paratroops: 123, reduced to 104 if a comfort pallet is included. Litters: up to eighty maximum, depending on configuration used. Passenger aft-facing 16G seat capacity: 138, reduced to 120 if comfort pallet is onboard. Cargo compartment is pressurized (to sea level up to 21,000 feet; to an 8,000-foot altitude pressure at 40,000 feet), allowing people to ride in comfort as well as transporting hardware of any sort. Cabin at forward end contains a hot-meal galley, sink, and refrigerator, with adjacent lavatory. Seats, litters, and paratroop configurations are unlimited due to quick-attach fittings. Oxygen is permanently installed and capable of being provided throughout the cargo cabin for medical use or if cabin depressurizes. Carriage of Minuteman series ICBM shipping and storage container ballistic missile (SSCBM) containers by straight-in loading method. Can be moved up to 2,300 nm. Can also carry any type of tactical missiles that require overseas deployment. The all-position ramp can be lowered to a horizontal angle for truck, K-series loader, or dock straight-in loading. Can be lowered to ground level, providing a gentle (11 degree) sloped access. Petal doors can be opened to 55 degrees for airdrops, and opened further to 80 degrees for loading of ground cargo. Using standard 463L pallets, up to ten can be carried. Up to eight 8 × 8 × 10 cargo vans can be stored sideways. Paratroops can exit the aircraft in-flight by means of jumping off the opened cargo ramp, or by using the two opened troop doors. An air deflector is deployed into the airstream to lessen aerodynamic buffeting on troops that are egressing the plane. Total cargo usable volume including the ramp is 7,340 cubic feet. The clear cubic footage is 6,524 cubic feet, and the palletized volume is 5,284 cubic feet.

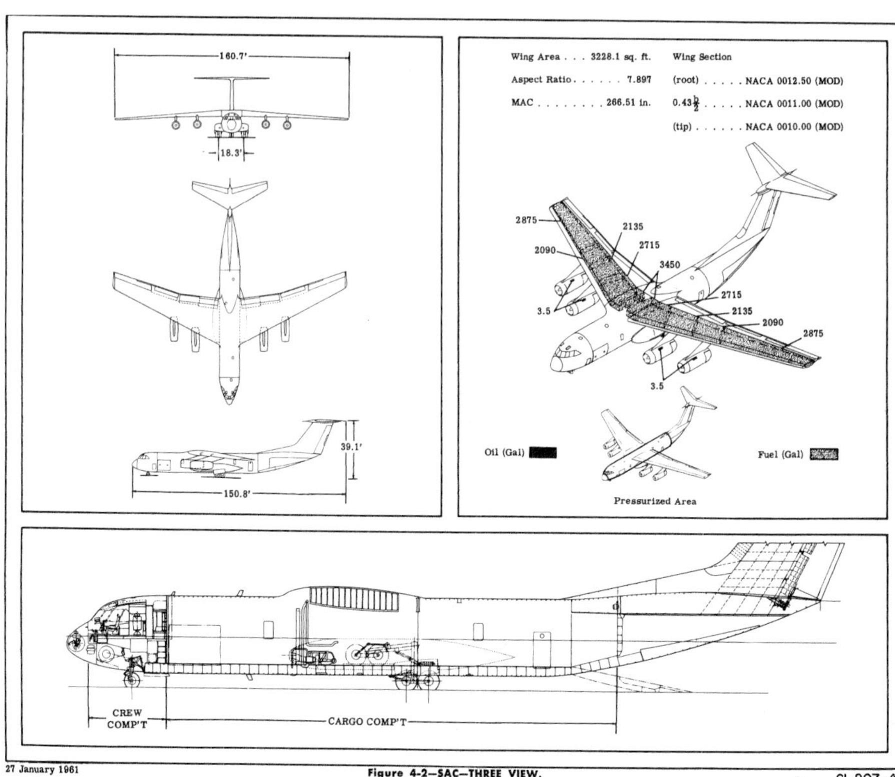

27 January 1961

Figure 4-2—SAC—THREE VIEW.

2 GL 207-45

POWER PLANT

No. & Model . . . . (4) JT3D-4
Mfr. . . . . Pratt & Whitney
Engine Spec No. . . 1761 Date 8-25-60
Type . . . . Axial Flow Turbofan
Length . . . . 136.32"
Diameter . . . . 53.14"
Weight (dry) . . . . 4170 lb
Tail Pipe . . . . Fixed Area

ENGINE RATINGS

S.L. STATIC		RPM	MIN
T.O.	18,000	---	5
Mil:	17,000	---	30
Nor:	16,400	---	Cont

DIMENSIONS

Wing
Span . . . . 160.67'
Incidence (root) . . . . 4.5°
(tip) . . . . 0°
Dihedral (.25c) . . . . -1° 15'
Sweepback (.25c) . . . . 25°
Length . . . . 150.813'
Height . . . . 39.115'
Tread . . . . 18.34'

Mission and Description

Navy Equivalent: None   Mfr's Model: GL-207-45

The principal mission of the GL-207-45 is to provide a rapid, reliable, and efficient means of airlifting combat or support units of all services under general or limited emergency conditions, military logistics supplies or commercial cargo, and mail.

The normal crew consists of pilot, co-pilot, navigator and systems engineer.

Features include integral ramp-pressure bulkhead and cargo door, crew and cargo compartment pressurization, ground and in-flight air conditioning, thermal de-icing for wing and empennage leading edges, single point refueling, palletized cargo loading, and provisions for air-dropping personnel and cargo.

Development

Date of Contract . . . . NA
First Flight . . . . NA
First Acceptance . . . . NA
First Service . . . . NA

GENERAL

CARGO
Max Cargo - See "Payload - Distance" graph, page 5.

CLEARANCES
Main Compartment:
Length . . . . 70.0'
Width . . . . 10.27'
Height . . . . 9.1'
Main Loading Door:
Width . . . . 10.27'
Height . . . . 9.10'
Height from Ground . . . . . . 4.16'
Side Cargo Door:
Width . . . . 9.10'
Height . . . . 6.5'
Height from Ground . . . . . . 4.16'
Paratroop Door:
Width . . . . 3.0'
Height . . . . 6.0'
Height from Ground . . . . . . 4.16'

CAPACITIES
Main Compartment . . . . . 6531 cu ft
Cargo Floor:
Single Axle Load . . . . 20,000 lb
Capacity . . . . 300 lb/sq ft
Ramp:
Single Axle Load. . . . . . 20,000 lb
Capacity . . . . 300 lb/sq ft

MISCELLANEOUS
Ramp (type) . . . . Integral
Ramp Incline . . . . 11°

PERSONNEL
Crew (normal) 4
Troops (max) 95
or
Paratroops (max) 74
or
Litters (max). 72
plus
Attendants 8

WEIGHTS

Loading	LB	L. F.
Empty	118,076 (E)	
Basic	126,070 (E)	
Design	315,000	2.5
Combat	*155,100	
Max T. O. (nor)	315,000	2.5
Max Land	+315,000	2.5

(E) Estimated
* For Basic Mission
\+ Limited by gear strength 6 fps sinking speed.

FUEL

Location	No. Tanks	Gal
Wing, inbd	2	4270
Wing, outbd	2	4180
Wing, inbd aux. . .	2	5430
Wing, outbd aux . .	2	5750
Wing, center. . . .	1	3450
	Total	± 23,080

Grade JP-4
Specification. MIL-F-5624E

OIL

Nacelles . . . . . . 4 . . . . (tot) 14
Specification. MIL-L-7808D

± Limited by structure to weight and disposition shown.

ELECTRONICS

UHF Communication (2) . AN/ARC-50
VHF Communication (2) . . . VHF-101
H.F. Communication (2) . . . HF-102
AGA SC/SELCAL
Radio Set (Tacan) (2). . . AN/ARN-52
Astro-Navigation Set. . . . AN/APN-1
Radio Compass (2) DF-202
Weather-Navigation Radar . . RDR/1D
Doppler Radar AN/APN-501
Altimeter. AN/APN-141
IFF AN/APX-46
Gyro-Stabilized
Compass (2) . . . . . . Sperry C-11
Inertial Platform. . . . Litton LN 2C
Digital Computer. AN/ASN-24

27 January 1961

Figure 4-3—SAC—INFORMATION BLOCKS

3 GL 207-45

Two pages from the Standard Aircraft Characteristics (SAC) description by Lockheed. The document, dated January 27, 1961, shows data pertaining to C-141A fuel tank locations in the wings, fuselage compartments and overall dimensions, engine, mission, general cargo/capacity specs, weight figures, fuel amounts, and electronic hardware information. All provide a basic outline of the Starlifter design and what it is made to do. *Lockheed*

C-141A, 61-2775/MSN 6001, while in the process of FAA certification. Of interest is the nose boom and opened normal crew entry door. The Starlifter (sometimes spelled as StarLifter) can best be explained in layout this way—a large transport plane, with a shoulder-mounted, swept wing and four wing-mounted, podded jet engines with a high T-tail arrangement. After testing, the nose probe was removed. The designation also changed, to NC-141A, along with the second, third, and fifth examples built, to reflect their status as permanent test bed airframes not intended for fleet use. Large flaps, used for slow-speed handling while landing, are also seen lowered. *National Archives via Dennis Jenkins*

C-141 Starlifters, 61-2775/MSN 6001, 61-2776/MSN 6002, and 61-2777/MSN 6003, on the ramp at Marietta. Six-digit BuNos are assigned by the government once the plane is accepted for use. The first two digits signify the fiscal year the item was accepted. The MSN is assigned by the company making the plane. The number is referred to when having to do depot maintenance or upgrade projects. Plane in foreground has upper wing spoilers deployed, nose boom protected with red covers, and a maintenance stand, to prevent the probe being bent or broken off by ramp vehicles moving about. Yellow items distributed about the ramp section are aerospace ground equipment (AGE) hardware, all used to service the aircraft, preparing them for flight or providing power, hydraulic pressure, or access during maintenance downtimes. *Lockheed*

The C-141 was the first jet transport that Army paratroops jumped from, in August 1965. This was not a primary requirement when SOR-182 was written, but the aircraft was easily adapted to roles involving mass airborne assault scenarios that included paratroopers to deploy. They exited the plane for the most part through the aft pair of troop doors. Men with gear pose beside 61-2779, the fifth plane built. *Lockheed*

C-141A, 63-8077, the eighth plane built, shows off its ability to provide straight-in loading as a K loader aligns itself with the lowered ramp. Seen here, the ramp is set level. It can be lowered to ground level, so that vehicles can drive into the cargo section. The load in this photo can be pushed manually into the plane, then it can either be pushed in farther if the cargo floor panels have rollers, or it can be pulled in with a cargo hoist. The load is lashed down and checked frequently in the course of a flight until unloaded. *Lockheed*

Features not adopted on the C-141A were forward/aft sliding troop doors, later made to slide up inside the cabin when opened. Deleting side loading door saved 200 pounds in empty weight. Aerial refueling was to be seen later in the C-141B stretch modification. *Lockheed*

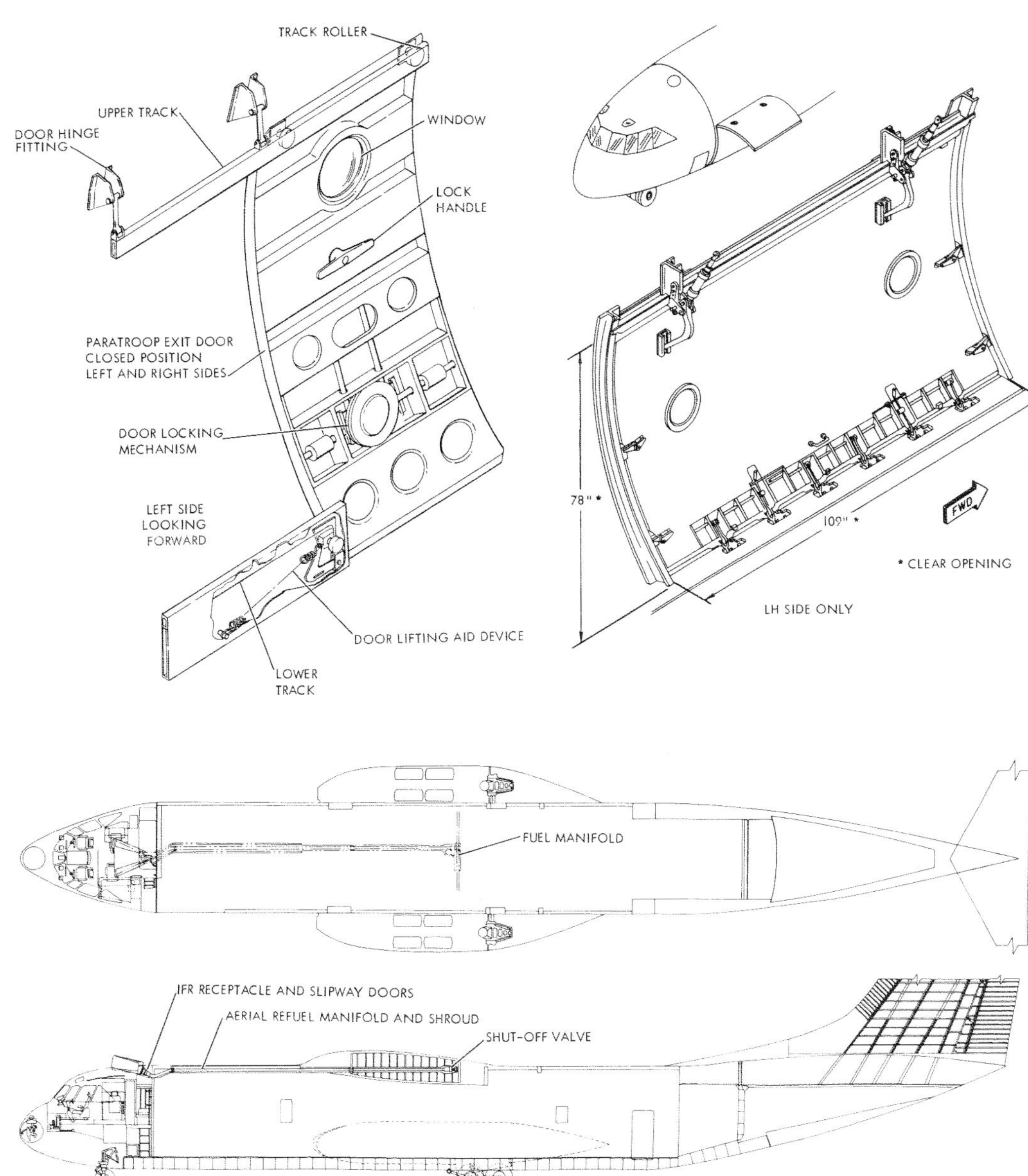

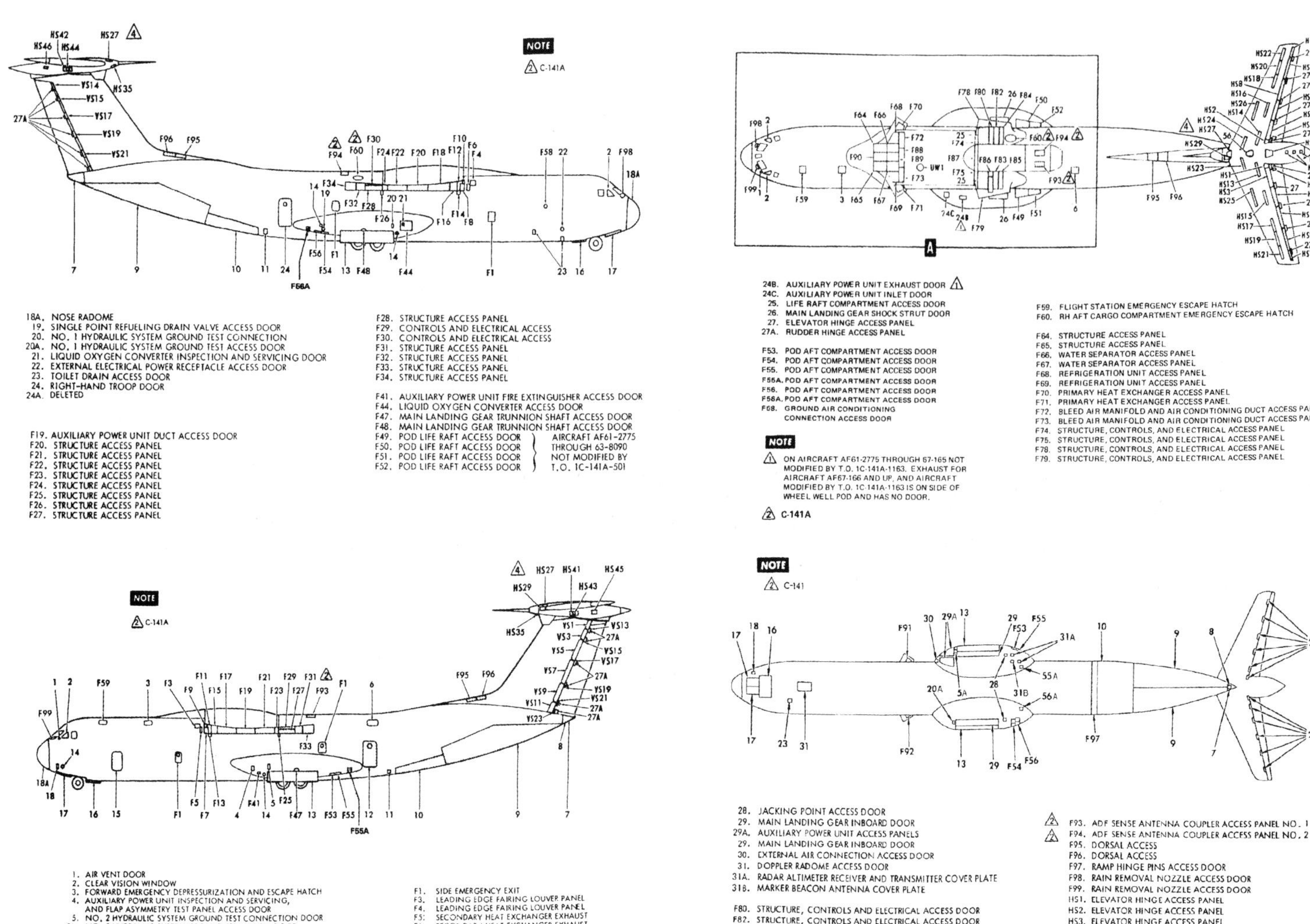

C-141A left, right, top, and bottom access panel locations. One SOR-182 line item that was emphasized was ease of maintenance. Getting to areas to view, troubleshoot, and repair components is therefore made much easier by having panels placed strategically all over the airframe. *USAF*

October 19, 1964, Tinker AFB, as C-141A, 63-8078, the "Spirit of Oklahoma City," is welcomed to the base with an Air Force band, spectators, and aircrew at attention.

The Wright Flyer was a replica, but the Starlifter is 61-2775, first off the production line. The C-141 was on the way to a long and fruitful career in USAF fleet service.

With the C-130 Hercules, Lockheed already had innovative and useful design features. The C-141 adopted many of these in an all-jet transport. The C-5A Galaxy added size and modern engines, as well as front/rear loading and greater range. *All photos this page, Lockheed*

One aircraft was built as the Model 300/N4141A, intended to fill civil airline orders. In the end, despite some interest, none were bought, so Lockheed used it for a time as a company demonstrator, eventually bailed to NASA as N714NA (NASA 714). It was modified to carry the Kuiper telescope for high-altitude astronomical observations. The aircraft is no longer in use. *Lockheed*

C-141A pilot side console, circuit breaker panel to left. Wheel with knob is the tiller, used to steer the nosewheels while taxiing on the ramp. *Mark Aldrich collection*

C-141A instrument panel. Flight gauges are in pilot and copilot line of sight. Engine gauges are clustered at center, with throttle quadrant below. *Mark Aldrich collection*

Navigator position behind pilot. Radar screen has hood, with LORAN-C avionics below it. Main panel has radar and navigation settings. *Lockheed*

Copilot side view of seat, control yoke, side console, and flight instruments. Throttles are set at the IDLE position. Pedals under panel operate rudder/brakes. *Mark Aldrich collection*

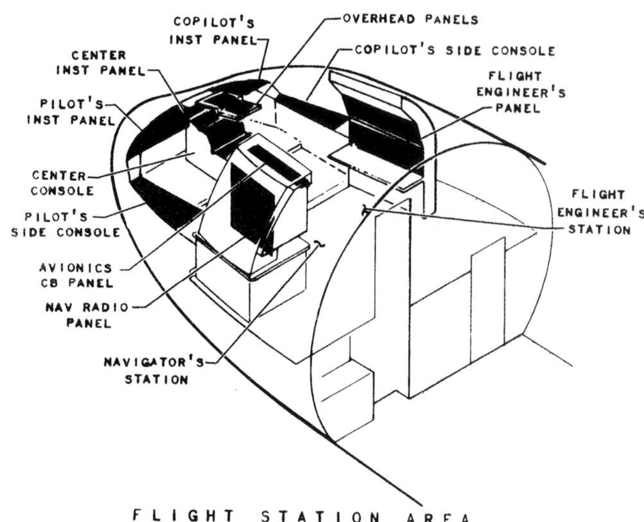

Aircrew positions in a C-141 cockpit. *USAF*

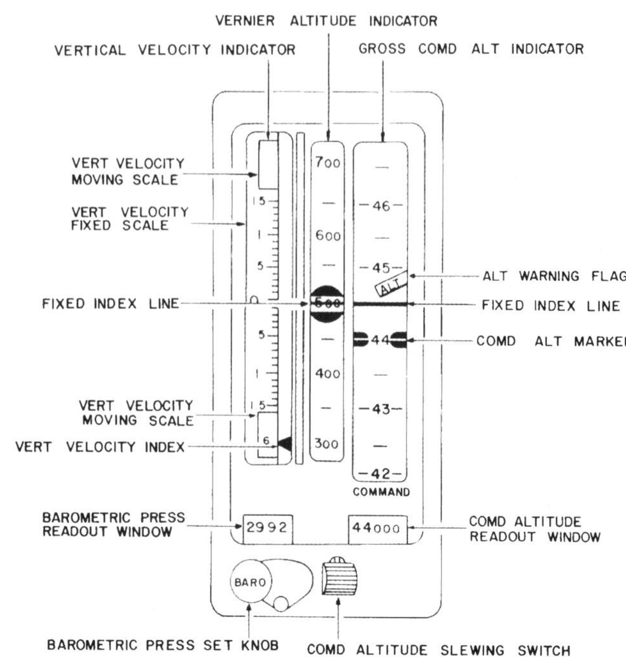

Altitude and vertical-velocity (VVI) indicator labeled. *USAF*

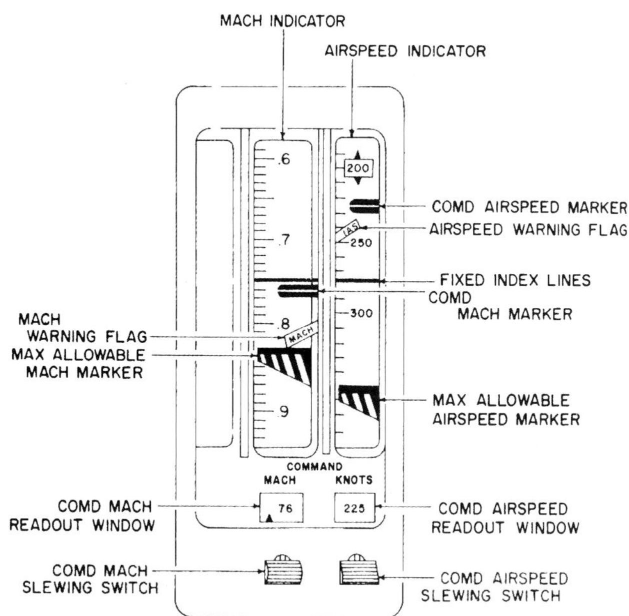

Vertical tape gauges on C-141 instrument panel. *USAF*

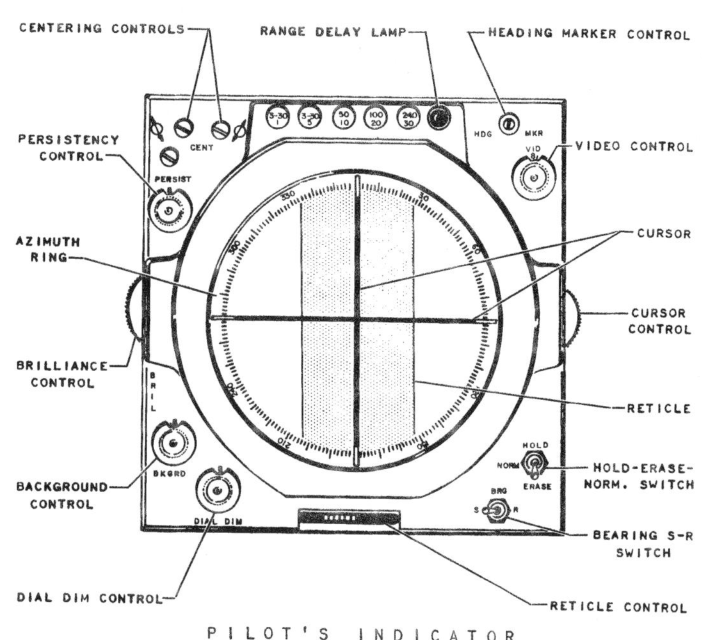

APN-59 radar display for pilot, labeled. *USAF*

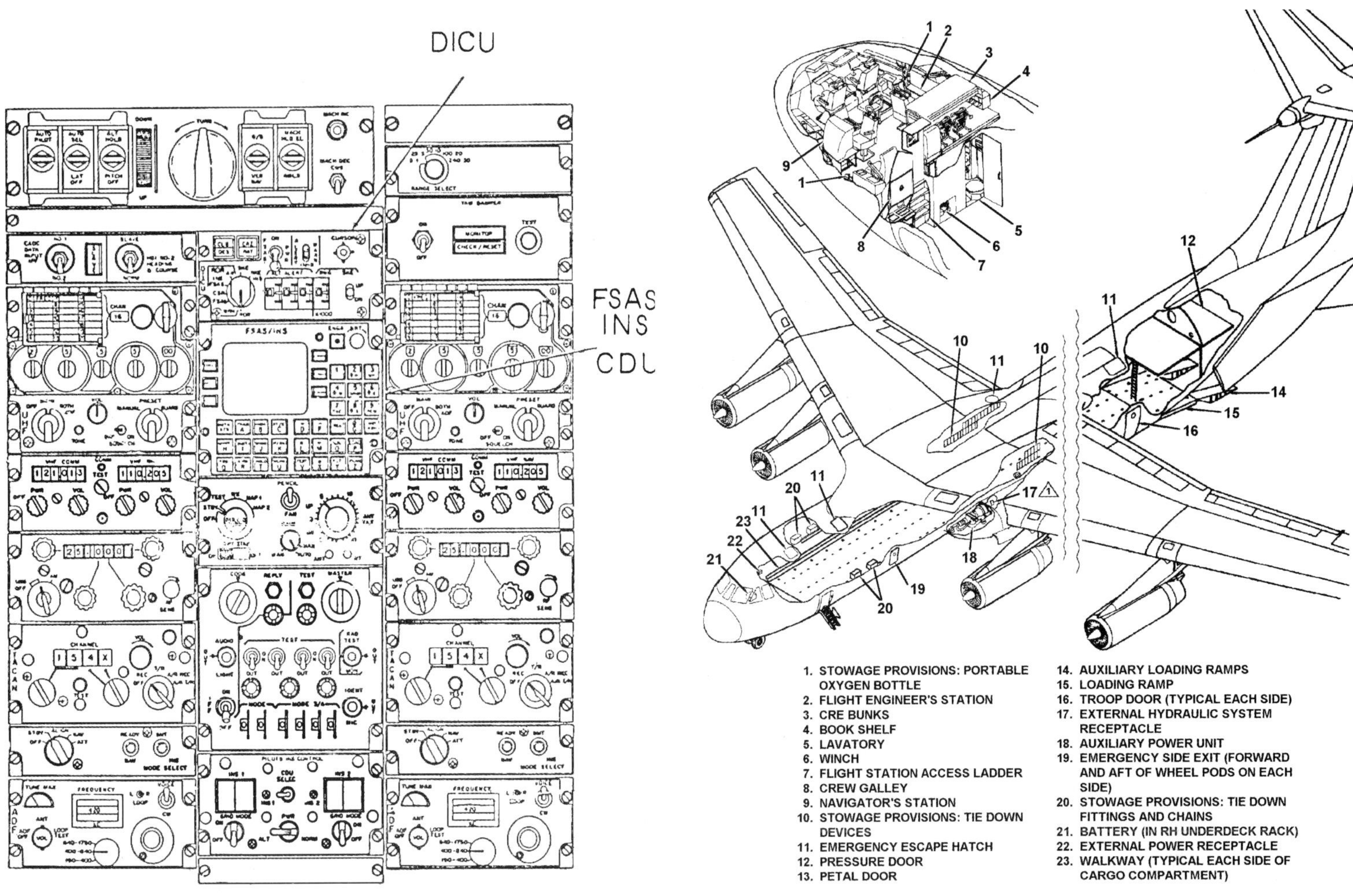

C-141 center pedestal, located between the pilot and copilot seats. Labeled panels are the display indicator control unit (DICU) and the fuel savings advisory system (FSAS). INS refers to the inertial navigation system. CDU label is used to highlight this as a control display unit. The three rows of panels include operating systems such as radios, the automatic direction finder (ADF), the tactical control and navigation (TACAN), and the identification friend or foe (IFF). Diagram to right shows C-141A interior general layout of the forward section and cargo bay. The auxiliary power unit (APU) is shown in the forward part of the port main gear fairing. *USAF*

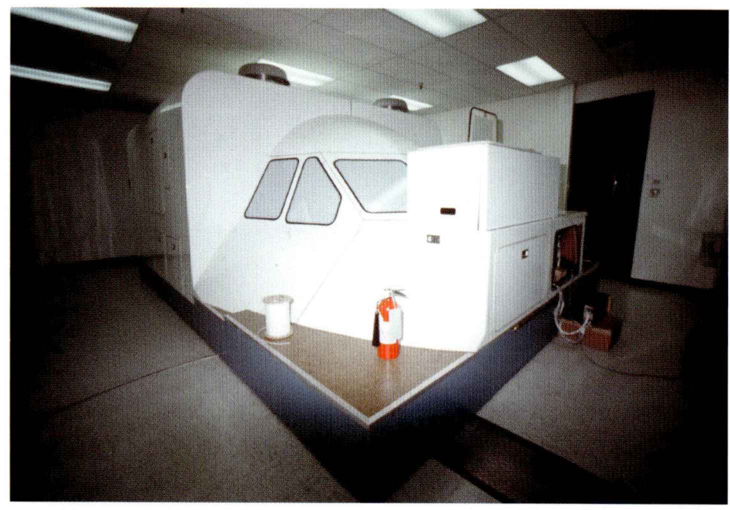

Inside a C-141C simulator at the Air Mobility Command Museum at Dover AFB, Delaware. Keyboard to input flight scenarios is to right. *Author*

Two views of a C-141 cockpit simulator, used to train aircrews in phases of flight such as takeoff, cruise, how to react to emergencies, and landing. *James E. Hayes / USAF*

MAC airlift training center at Altus AFB in Oklahoma, where C-141 and C-5 advanced instruction is carried out. Visual aids simplify systems details. Airman Magazine*, June 1988*

In the left photo, brand-new TF33-P-7 engines are moved along an assembly line. The air intakes are at bottom. Components are added at each phase of the assembly process. Many wires, hydraulic lines, and sensors that will eventually be connected to cockpit gauges are already installed. The image on the right shows workers cleaning the inside surface of a petal door. While cleaning, they look out for any imperfections, gouges, scrapes, or dents that may require further finishing. The doors are designed to withstand buffeting and vibration effects while being opened and closed as the plane is in flight. *Mark Aldrich collection*

The vertical tail fin assembly. Top is in background, where the all-moving horizontal surfaces will be joined. *All images this page, Mark Aldrich collection*

In the final stages of assembly, a horizontal stabilizer gets inspected. A technician vacuums out the void spaces before panels get fastened into place.

A C-141 tail fin of vertical and horizontal flight surfaces is mated together. The rudder will be hinged to the trailing-edge-facing camera.

Horizontal stabilizer in foreground has most of its access panels fastened to upper surface. Wires are grouped and routed along leading edge.

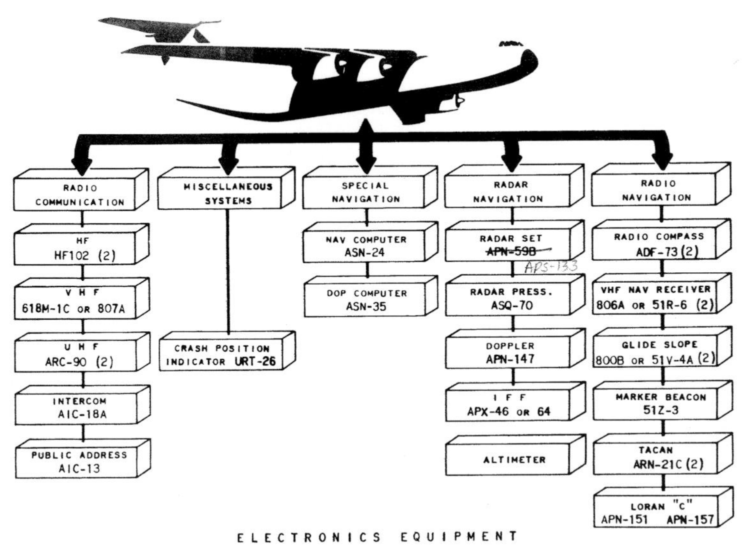

A listing of the electronic components installed in the C-141A. *All images this page, USAF*

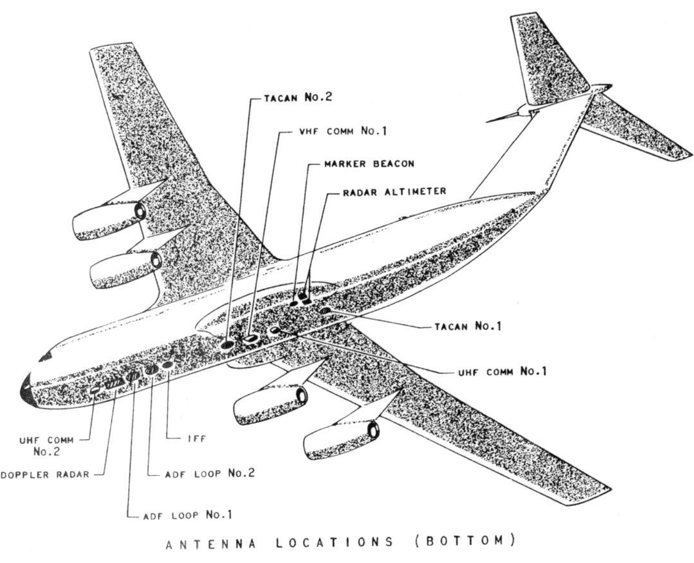

C-141A antenna locations, bottom view

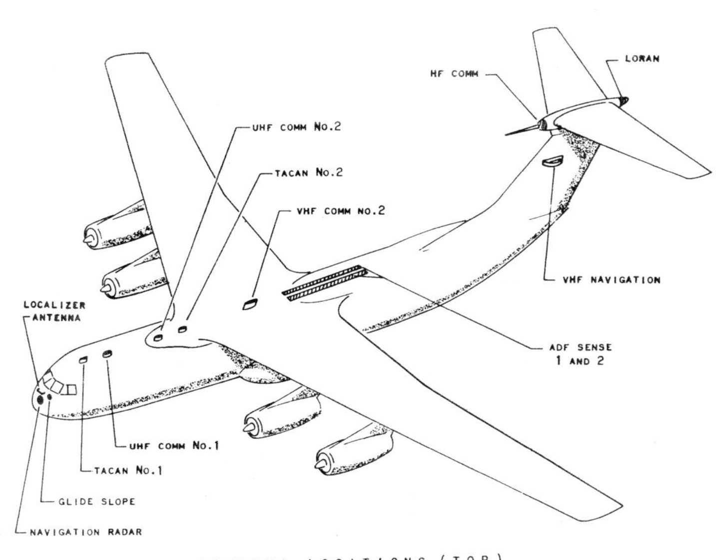

C-141A antennas, shown on the aircraft, top view

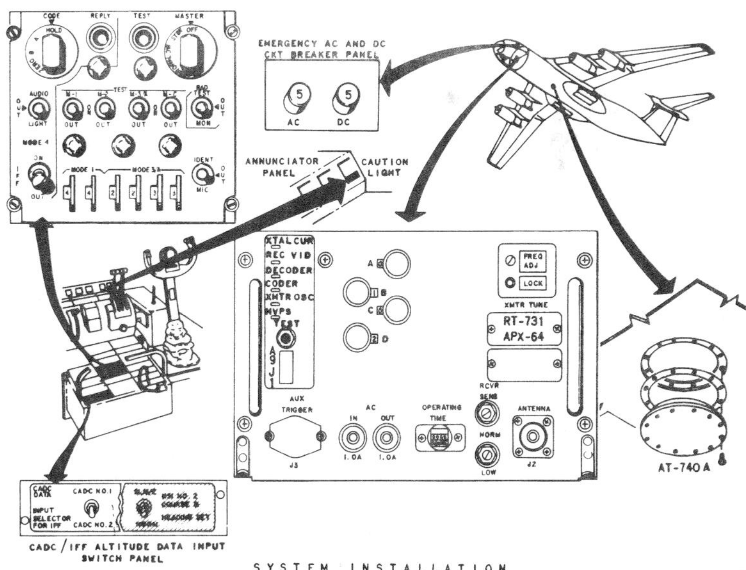

The IFF system, showing antenna, supporting hardware, and panel locations

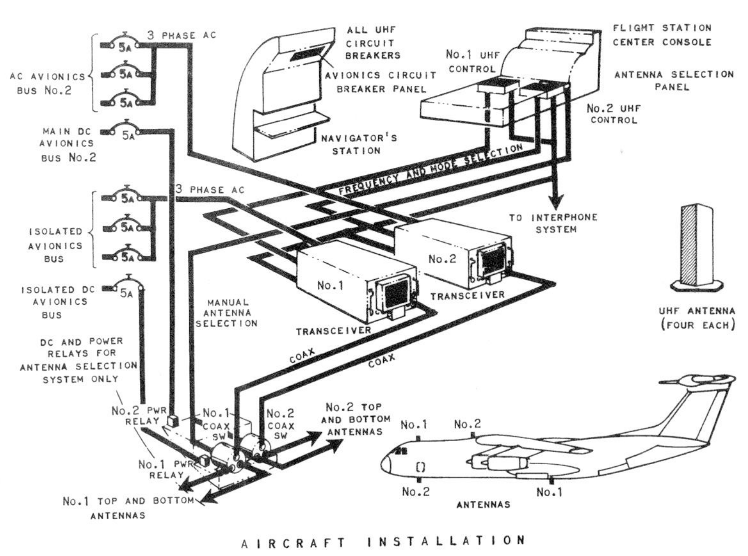

C-141A UHF radio communications. *All images this page, USAF*

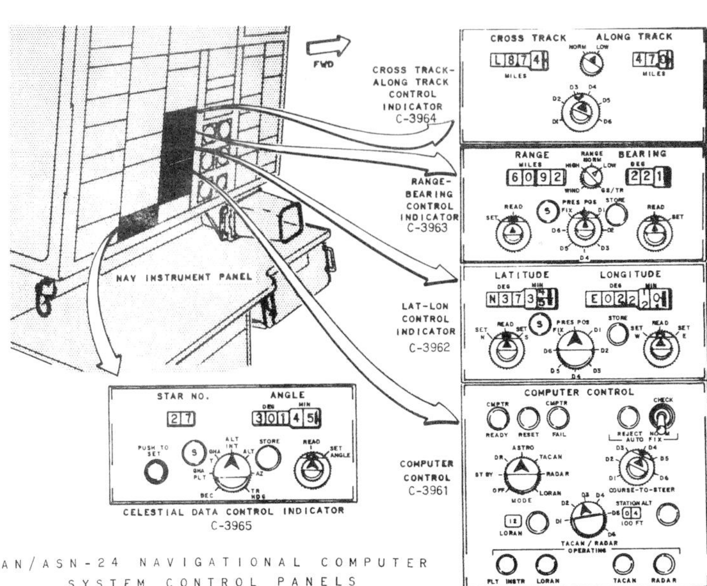

ASN-24 navigation system components on panel

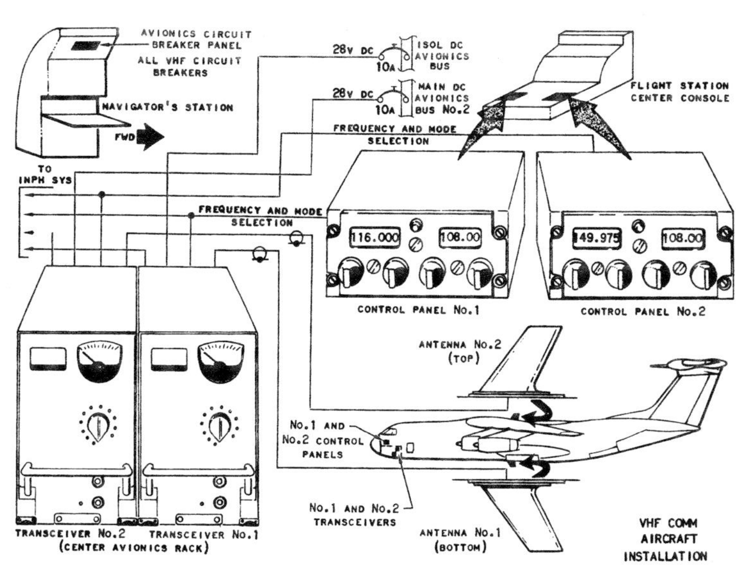

The VHF setup on the C-141A

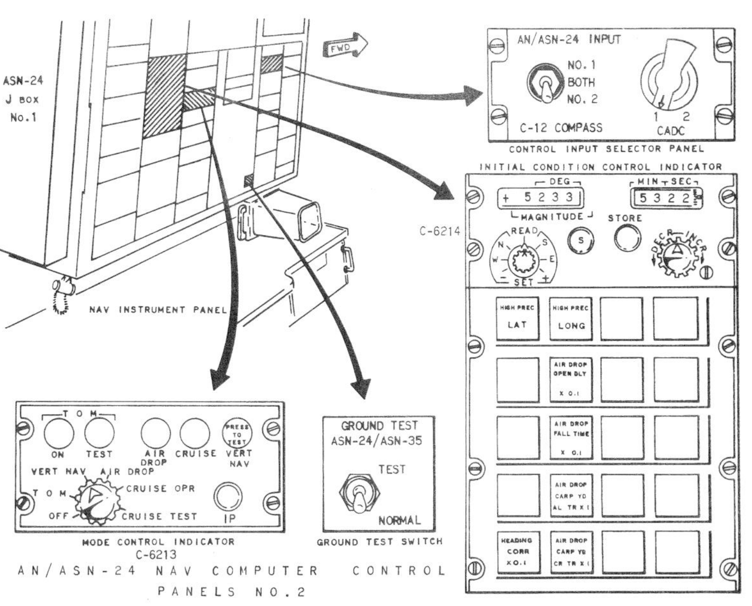

ASN-24 panels, with their placement on console shown

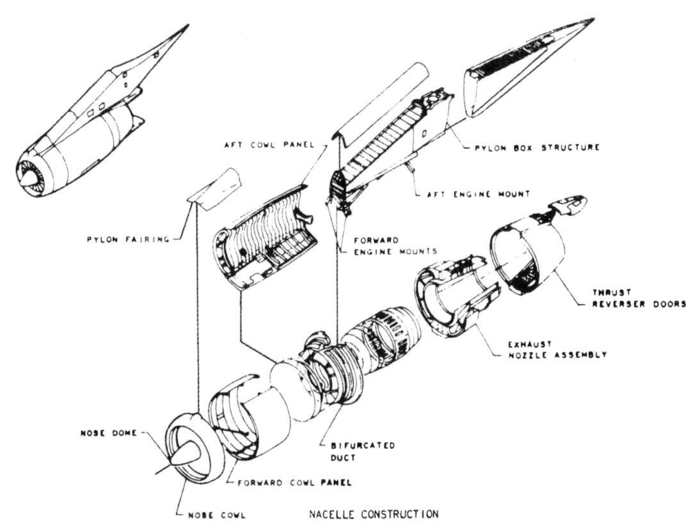

TF33-P-7 engines, four of which power the C-141

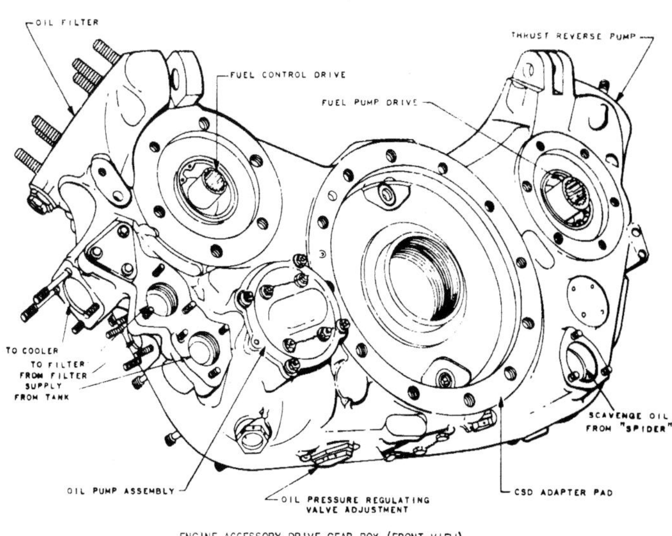

Accessory drive gearbox, attached to each engine

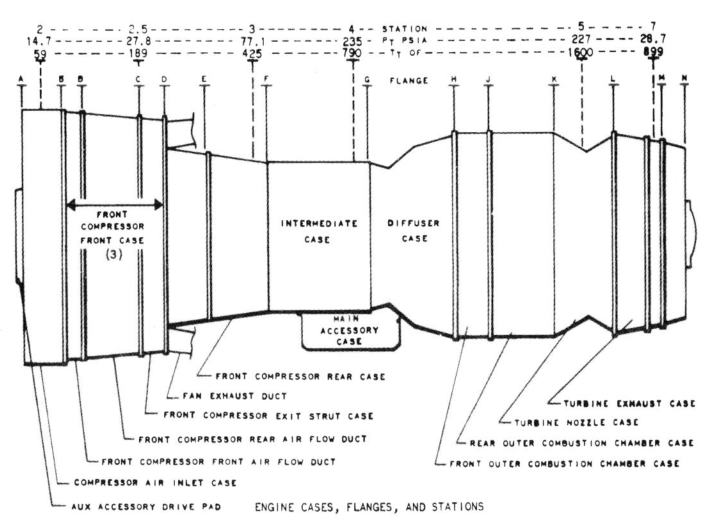

Engine major components. *All images this page, USAF*

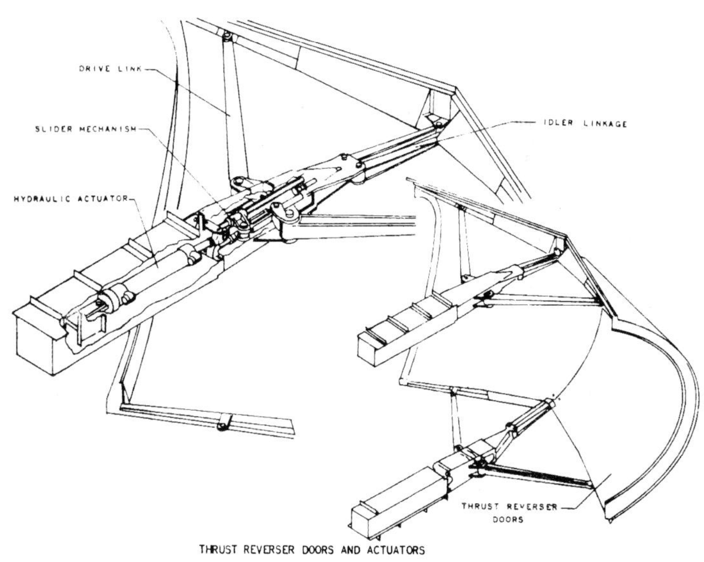

Thrust reversers, used to slow plane down on landing

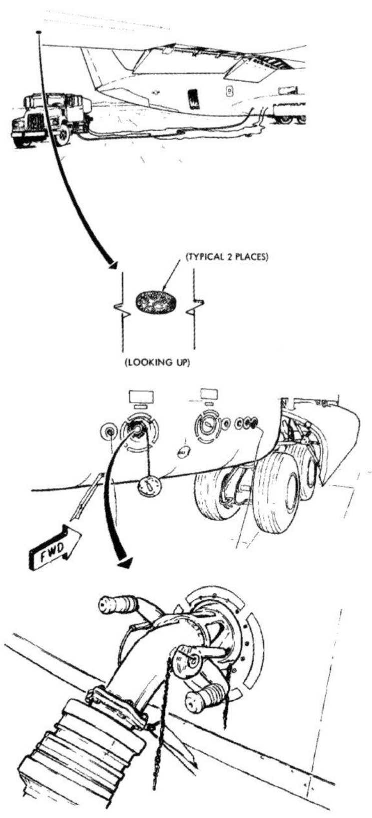

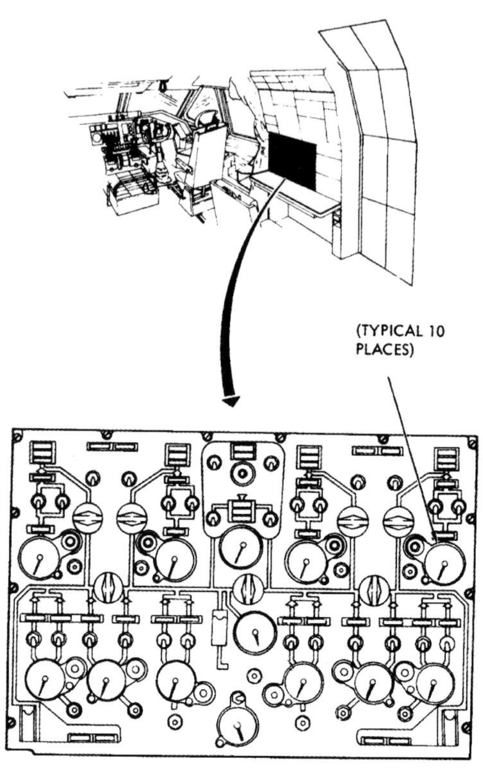

SPR Receptacle

Fuel Servicing Panel

C-141 fuel system, showing wing vents (*at top left*) and single-point-refueling SPR pressure hose connection in left main wheel fairing. *At top right*, flight engineer's panel in cockpit and fuel panel location. *At lower right*, fuel panel layout of knobs and gauges. *USAF*

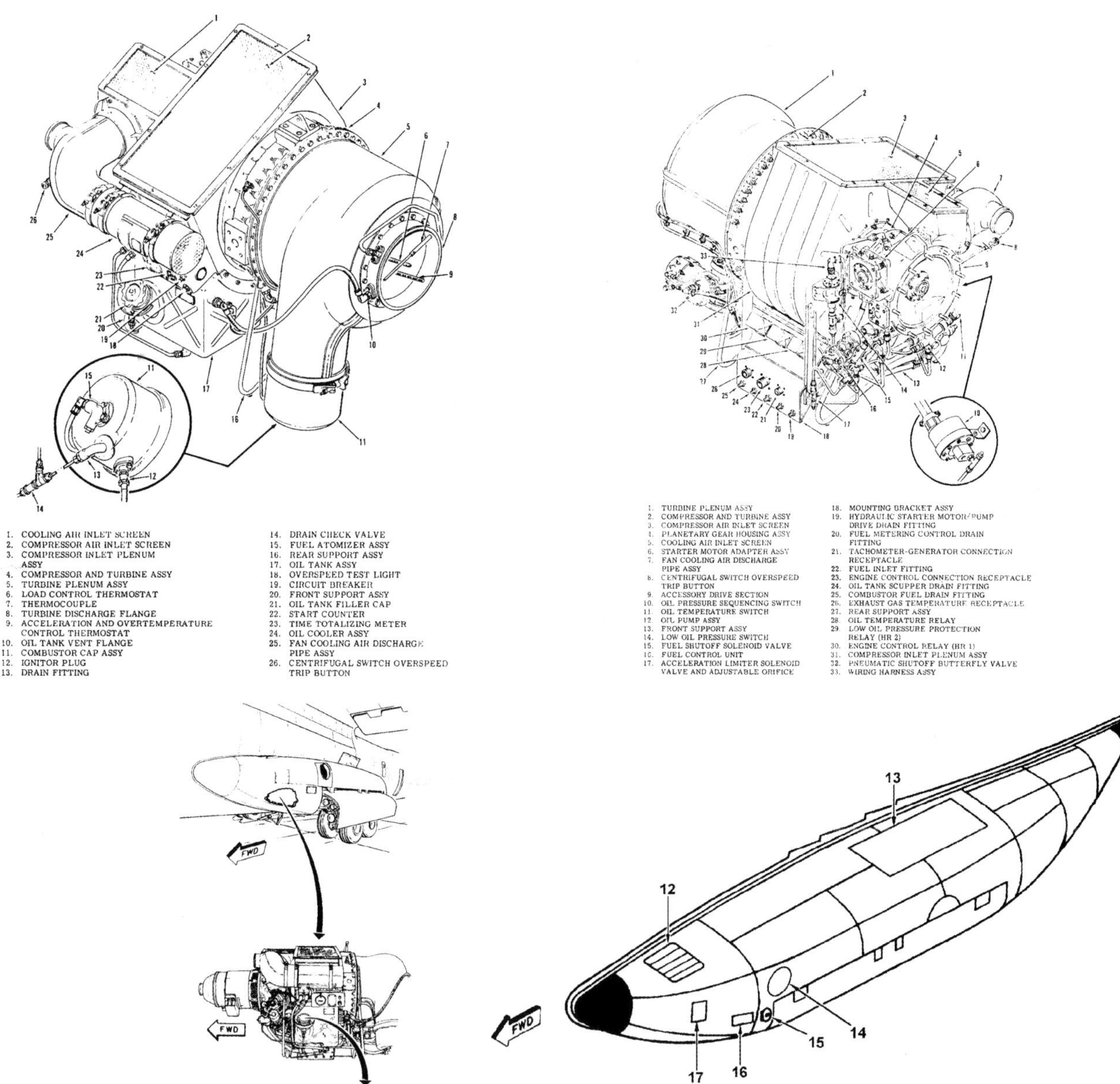

Auxiliary power unit (APU) features. Two top views show labeled items on unit. *At lower left*, oil check/servicing location; *at lower right*, port (left) main gear fairing. Item 12 is APU air inlet door, item 14 is APU exhaust outlet, item 16 is APU fire door, and item 17 is APU inspection/servicing access door. *USAF*

C-141A, 61-2776, poses at Edwards AFB, California, in May 1965 with one of the North American XB-70A Valkyrie supersonic bombers, and the YF-12A, interceptor variant of the better-known SR-71A reconnaissance aircraft. *Marty Isham collection via Dennis Jenkins*

At Edwards, C-141s were put through their paces, testing their abilities in flight and in ground operations, such as this landing with thrust reversers engaged and full antiskid braking applied to the main wheels. 63-8075 performs an emergency stop as a T-38 chase aircraft takes in an elevated view of the scene. *USAF*

Not a scene inside some drafty hangar at a Far North remote airbase, but in Florida, where C-141A, 63-8076, was placed inside the McKinley Climatic Laboratory at Eglin AFB for cold-weather testing. In frigid conditions, many items such as fueling, engine runs, cabin environmental-control system (ECS) tests, oil viscosity in components, and support equipment operation all were examined and verified before use of the plane in such extreme conditions. Other weather extremes can also be duplicated, such as heat and rain. *USAF*

One SOR-182 item was for the new cargo plane to carry an ICBM missile. This photo shows C-141A, 65-0235, in the process of loading a container, inside which is a Minuteman I. Moving missiles by air to facilities where the missile silos are located enhanced mobility and security, as opposed to long trips on roads or by railway transport. *USAF*

Another role taken on by Starlifters was that of aeromedical evacuation. The cargo cabin could be adapted for litter accommodation for up to forty-two patients, along with the medical staff and equipment necessary to attend to them. Petal doors shield the buses or ambulances from wind, heat/cold, and rain, while loading and unloading occurs as a side benefit. *USAF*

SSgt. Cathy Coleman, of the 31st Aeromedical Evacuation Squadron, greets a patient before being loaded aboard. *William W. Magel / USAF*

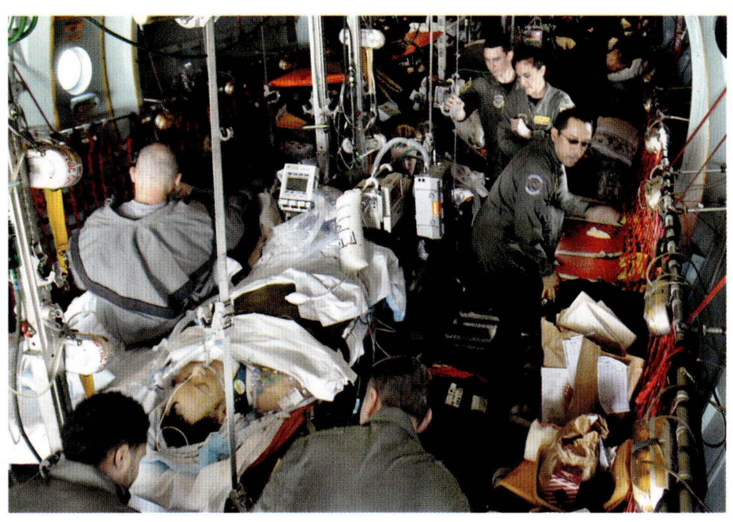

A C-141 configured as a medevac loads up patients in Germany as medics prepare equipment needed for the flight ahead. *Paul Minert collection*

A simulated victim during the Reforger '81 drill is brought aboard directly from an ambulance at Ramstein AB. *Bill Thompson / USAF*

Medics of the 2nd Aeromedical Evacuation Squadron rest on a flight to Pakistan to pick up Afghan freedom fighters for further treatment in the USA. *Nolan/USAF*

C-141 Timeline, 1965–1968

March 23, 1965: The Air Force signs a contract with Lockheed for eighteen more C-141s.

April 23: C-141A 63-8088, the "Golden Bear," is delivered to Travis AFB in California. The same month, routine flights to and from the Southeast Asia theater begin, supporting the Vietnam Conflict commitment.

May 25: The "Golden Bear" flies from Travis to Japan in nine hours and twenty minutes.

June 10–21: The 26th Paris Air Show hosts a C-141 placed on display.

August 5: Airlift to Saigon, Republic of South Vietnam, begins. First flight delivers a load of 50,000 pounds of cargo. Flight from Tinker AFB to Saigon takes eighteen hours and fifteen minutes.

August 14: Charleston AFB receives its first C-141.

August 18: First C-141 is flown to Dover AFB.

December 31: A total of sixty-five Starlifters had been delivered to MATS bases.

December into January 1966: Operation Blue involves C-133s and C-141s, delivering 2,952 troops and 4,749 tons of cargo from Honolulu, Hawaii, to Pleiku, South Vietnam.

January 31: First C-141 able to haul a containerized Minuteman ICBM missile is delivered.

April 1: C-141s start regular embassy support flights and aeromedical evacuation taskings in the European theater.

April 14: First low-level combat airdrop occurs at Fort Bragg, North Carolina.

April 21–22: In a test exercise, a Minuteman missile container is loaded, then unloaded.

April 27: The 100th C-141 is taken by the Air Force, and tests of an all-weather landing system (AWLS) are begun.

C-141A, 63-8075, with the flight milestone of 2,500 hours in eleven months on the nose.
Warren Bodie via Bob Esposito collection

May 15: The largest airdrop since the Second World War, Operation Air Drop III, takes place. Airlifters of all types, totaling 144 aircraft from twenty-eight units, deploy 6,000 troops of the 82nd Airborne Division and the United Kingdom's 5th Airborne Brigade to three drop zones at Fort Bragg.

June 30: The Air Force contracts for 134 more C-141s, bringing the total production run up to 284 planes.

July: A nonstop flight from Tinker AFB to Yokota AB in Japan takes twelve hours and fifty minutes. Distance covered is 5,948 miles.

August 5: McChord AFB, Washington, gets its first C-141, BuNo 65-0277. Assigned to 62nd Military Airlift Wing (MAW), 8th Military Airlift Squadron (MAS). In September, a C-141A, 65-0281, is destroyed at the base by an early-morning fire.

November 14: A C-141A, commanded by Capt. Howard Geddes, of the 86th MAS, Travis AFB, lands on the ice pack at McMurdo Sound, Antarctica. Flight was from 2,200 miles away, originating in Christchurch, New Zealand. Thus began the Deep Freeze resupply flights.

December: Deliveries of Starlifters to the Air Force total 164 examples.

March 23, 1967: An A-6 Intruder landing at Da Nang AB, South Vietnam, strikes C-141A, 65-9407, of the 62nd MAW, as it is crossing a runway. A-6 aircrew and the C-141 loadmaster survive the incident.

April 12: A C-141 crashes into Cam Ranh Bay attempting to take off. Cause found to be aircraft flight surfaces not properly configured for departure.

June 30: C-141s in service amount to 220.

October: FAA approves AWLS as certified for use.

February 1968: The last C-141A is assembled and delivered.

March: C-141 program operations is transferred to Air Force Logistics Command (AFLC)

C-141A, 63-8088, the "Golden Bear," with power cart connected, a jet mechanic tending to number 3 engine. *Pete Bowers via Bob Esposito collection*

CHAPTER 2
NC-141A Test Beds

Four C-141As were redesignated as NC-141A (61-2775, 61-2776, 61-2777, and 61-2779), with the letter *C* being used as signifying a cargo aircraft, and the *N* denoting an aircraft devoted to permanent special test status. This indicates that the plane is not assigned to regular fleet service, but instead to commands that work in research and development and in making temporary modifications in the pursuit of work to support the regular forces. These aircraft are detailed in pictures and text in this chapter.

61-2775: Modified numerous times to serve in various test programs, evaluations, and system development projects. Before it was retired, it was used in Project Eclipse, by towing a Convair F-106 Delta Dart fighter in the air as an experiment to determine if an alternate method of launching a space shuttle was practical. Preserved at the Air Mobility Command Museum at Dover AFB, Delaware.

61-2776: Flown over 1,000 hours to destinations around the world, such as Alaska, Kwajalein Atoll, Guam, Brazil, and many stops in the European theater, to test electrically operated flight controls in the first regular use of the system. Tested C-5 Galaxy electronic hardware.

61-2777: Extensively altered, with the large "tail can" aft fairing containing space for the installation of self-defense systems. It also had a large port forward side fairing housing the SAPPHIRE radar. The starboard forward fuselage had antennas used in the integrated multifrequency radar test set (IMFRAD). Had "Desert Rat" nose art. Broken up and salvaged at the boneyard, Davis Monthan AFB, Arizona.

61-2779: Used as the advanced radar test bed (ARTB), with the nose radome modified from a blunt shape to a pointed style common to fighter airframes. It also was fitted with a chemical laser for use as the LIDS platform. Call sign was AGAR 14. It had "Against the Wind" nose art.

Lockheed proposed many other designs for C-141s, including an airborne warning-and-control system (AWACS) variant, an aerial-refueling tanker, a bomber, a minelayer, a missile launcher, search and rescue, a navigation trainer, and an electronics platform, among others.

61-2775 while in testing with nose boom; main gear tires have white stripes. *USAF*

NC-141A, 61-2775, at rest on permanent display at the Air Mobility Command Museum, located at Dover AFB, Delaware. Air intake at wing root is for cooling airflow into environmental-control system (ECS) compartment. White topsides reflect heat, helping keep interior temperatures low. *Author*

NC-141A, 61-2776, on the ramp at Marietta, Georgia, in February 1968. Radome has been replaced with one of greater internal volume during testing of electronic components for the C-5 Galaxy. Hardware under evaluation included the multimode radar, inertial Doppler navigation set, and compass equipment. Waving is technician S. D. Cash. *Lockheed photo via John Vadas*

NC-141A Starlifter, 61-2776, in March 1983 in the above image. ASD in white letters inside black tail fin stripe signifies that the plane is assigned to the Aeronautical Systems Division. Tail cone is slightly modified from the standard configuration, and lightning inverter is not present. In the right image, taken in August 1993, the ASD lettering is gone. Desert Rat artwork on left nose. Flag is above black tail fin stripe. *Both photos, Paul Minert collection*

September 1970 photo of NC-141A, 61-2777, at Wright-Patterson AFB. Aircraft is unpainted and modified with the aft fairing, and several side fairings are used in the testing of fuselage-mounted arrays of electronic detection gear. Lower picture shows 61-2777 in November 1981. Just visible under "US Air Force" are reinforcing doubler plates and covers where the integrated multifrequency radar antennas (IMFRAD) were attached. It was evaluated as a method of detecting hostile troops and equipment under jungle foliage, using three distinct frequencies. It could look below as well as perpendicular to the aircraft flight path. *Top photo, Paul Minert collection; bottom photo, Ronald McNeil*

NC-141A, 61-2777, taxis at Wright-Patterson AFB, Ohio, in 1976. Large fairing below tail fin, at times referred to as the tail can or beer can, is an aft-facing space where electronic components can be installed on the large, round section for airborne testing. The large forward fairing contains the synthetic aperture precision processor, high reliability (SAPPHIRE) radar. Hardware installed in the plane included the APD-10 radar unit (*shown*), a preprocessor unit, and a twenty-eight-track tape recorder. Aerodynamic evaluation of the radome was conducted, with fixes made, in 1975. Seventeen flights were made in 1976. Inset above shows the 3-frequency IMFRAD antennas which at one time were attached to the forward right-hand side of the fuselage. *Main photo J. W. Hawkins collection/Inset John Vadas collection*

Two photographs of NC-141A, 61-2777, in August 1975. In the left image, the "beer can" fairing with covered openings, into which sensors or electronic antenna elements can be inserted. The right picture, taken in early 1985, shows three radomes, part of the B-1B Lancer tail-warning addition of a Doppler radar set to the ALQ-161 electronic-warfare suite. In a series of background clutter tests, inert 2.75-inch folding-fin aerial rockets were fired to explore the sensors' capability to detect them. Flying hundreds of hours and expending hundreds of rockets, the system was finally declared ready in late 1986. Later use of the aircraft involved mounting missile approach warning-system sensors such as AAR-47 on the fairing to test their ability to detect missile motor exhaust plumes, with the firing of over eighty missiles in trajectories that allowed the sensors to view them. *Left photo, J. W. Hawkins collection; right photo, Mark Aldrich collection*

NC-141A, 61-2779, configured as the airborne radar test bed (ARTB). It was designed as a dedicated radar test platform, built under a USAF Aeronautical Systems Division contract with Lockheed Aeronautical Systems Company. The nose housing initially accommodated the APG-63 radar unit, but it could also fit the APG-70, APG-66, APG-68, or APQ-164 sets. The ARTB flew these radars in simulated electronic-warfare (EW) scenarios. *Paul Minert collection*

Modified nose on 61-2779. Last three of BuNo on side aft of radome bulkhead line. Yellow weights are used when aircraft is on jacks. *John Vadas*

ARTB being outfitted for radar installation. Ample space for avionics racks is behind the fiberglass radome. This provides maximum flexibility for testing. *Lockheed*

NC-141A is painted and has "Against the Wind" artwork. Small black pod contains a weather radar. Consoles inside record and monitor events. *John Vadas*

ARTB provided a dynamic method in which to test radars at many locations in real-world EW situations, recording and relaying results to ground stations. *Lockheed*

NC-141A, 61-2779, at Eglin AFB for vibration tests during its use as the laser IRCM demonstration system (LIDS) test bed. Sensors to measure vibrations included forty-six accelerometers and a single angular-rate sensor. Image is partially light-exposed but shows items highlighted during an underway test with petal doors open. Fire crash crew wearing green-tinted goggles was needed in case a hazardous-materials incident occurred from any chemical spill. A large white board off to right was used to calibrate the laser system. Laser components weighed 5,000 pounds. Ballast was positioned forward to compensate for the aft center-of-gravity condition. The airplane was flown in a low Q (150 knots / 24,000 feet) profile and at a high Q condition of 350 knots / 18,000 feet. *Author*

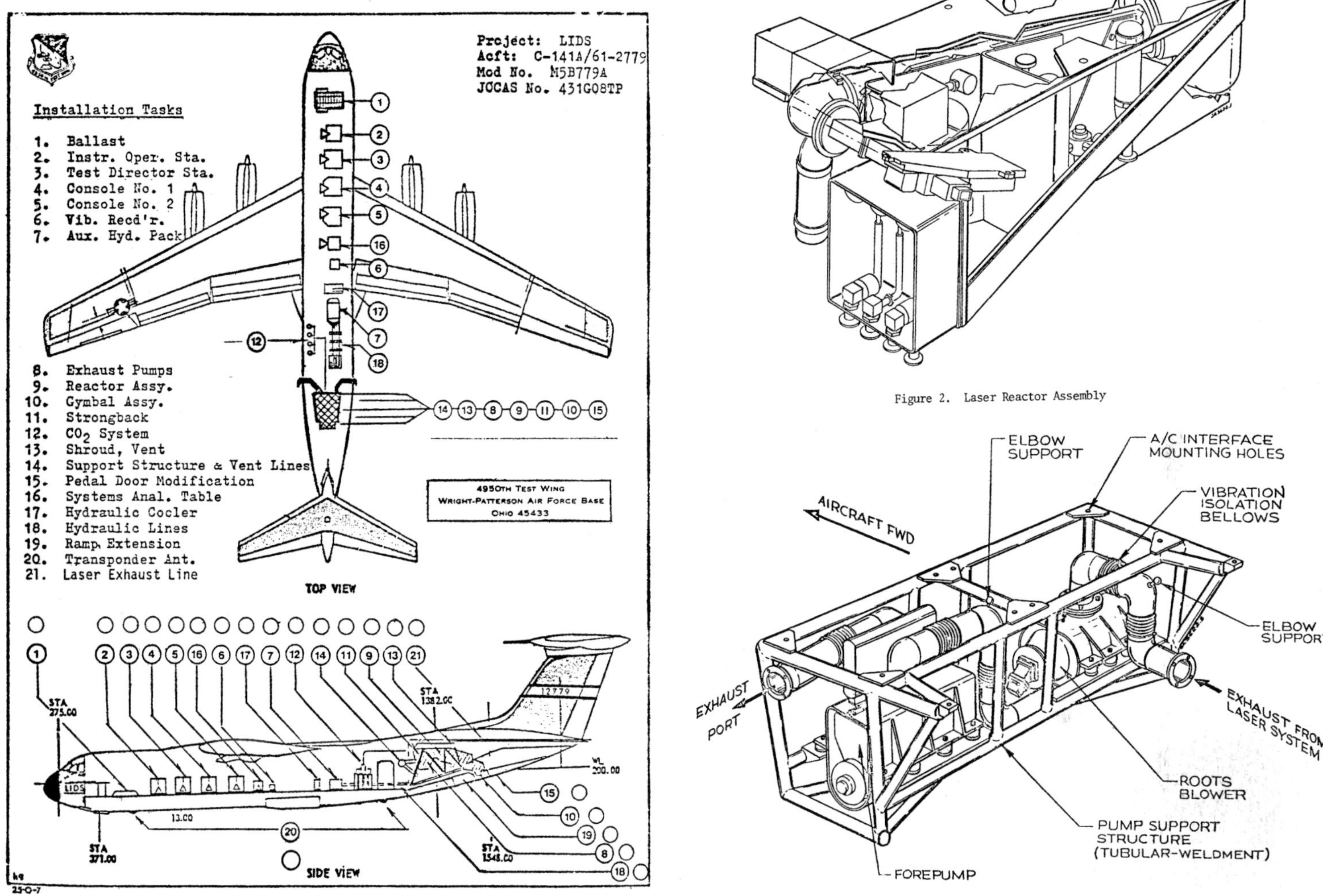

Figure 1. LIDS Profile Drawing

Figure 2. Laser Reactor Assembly

Figure 3. Laser Exhaust Assembly

NC-141A, 61-2779; layout for the LIDS installation is shown in the left image. Side view shows how far aft the laser is placed, necessitating the ballast noted as item 1. Various operator stations are labeled. Vent lines (#14) allow gases from chemical reaction to exit overboard. Two drawings at right show the top view of the laser reactor assembly, with the exhaust equipment shown in the lower rendering. *USAF*

CHAPTER 3

Vietnam and Antarctica

C-141s were put to use supporting the conflict in Southeast Asia soon after being introduced into MATS service. With their faster speeds and ability to load and unload quickly, a network was established, using MATS (renamed MAC on January 8, 1966) bases on both coasts of the continental United States, into bases in South Vietnam such as Tan Son Nhut, Da Nang, and Cam Ranh Bay, among other locations. En route stops included Alaska, Guam, Hawaii, Japan, the Philippines, and Wake Island. Travis AFB in California soon became the main hub for movements by air of men and cargo headed into the Southeast Asia theater. On August 3, 1965, the first C-141 flight from Travis to Vietnam occurred. Delivered to Tan Son Nhut during an eighteen-hour total time was 44,315 pounds of cargo, with stops on Wake and at Clark AB. October began the "Fast Fly" program, in order to meet the airlift demand. This was not enough, so as a boost to augment airlift capability, the "Red Ball Express" was added. In 1966, 9,353 tons of cargo and troops were delivered by air. This increased in 1967 to 19,318 tons. More bases were added as airlift destinations, such as Bien Hoa, Korat, Utapao, and Udorn in Thailand. As cargo was unloaded on airfield ramps at all these locations, smaller airlifters, such as C-7 Caribou, C-123 Provider, and C-130 Hercules transports, loaded up and delivered goods throughout the theater of conflict.

C-141s would shuttle about the region, from Japan to the Philippines and Vietnam, for six-to-ten-day shifts before heading back to Travis. On many occasions, once cargo and troops were unloaded, the plane would be reconfigured for a medevac flight, leaving for Japan and, from there, back to the US once patients were cleared to travel home. The faster airspeed of the Starlifter was credited as saving lives, as a result of shorter times in transit to medical facilities. The C-141 proved its worth time and again, such as in December 1965, when the 25th Infantry Division was airlifted from Hawaii to Pleiku. Moved by air were 2,952 men and 4,479 tons of cargo. In 1967, the 101st Airborne Division was airlifted to Bien Hoa AB in an operation involving the carriage of 10,024 paratroops and 5,337 tons of gear. The year also saw the most cargo volume moved, totaling 259,462 tons. In the wake of the Tet Offensive, 9,299 medevac patients were processed through Travis. In July 1968, C-141s flew a total of 4,535 soldiers and 480 tons of related equipment from Fort Carson, Colorado, to Bien Hoa. In 1969, the total was 1,259,462 individuals having passed through on their way to overseas locations. Sadly, the base mortuary recorded that 10,523 caskets had also been documented as delivered from Southeast Asia.

Beginning in 1966, C-141 support of Operation Deep Freeze (ODF) continued for thirty-nine years with no mishaps or anyone left behind at the National Science Foundation's McMurdo Station. Located on the southern tip of Ross Island, it has an airstrip 10,000 feet long, known as the Pegasus Runway, which consists of pressed/rolled snow on the permanent ice shelf, which moves 115 feet each year. On average, 200 flights were conducted each year, staging from Christchurch in New Zealand. On the way down to the station, there are no alternate landing fields an hour out from Antarctica. The distance one way is 2,076 miles, taking five hours to complete. The last C-141 flight (by 66-0152) was on February 4, 2005. In that year, a total of 1.7 million pounds and 3,000 passengers were moved to and from the Antarctic research station. The biggest challenge is weather related, as well as blowing snow, which can hide visual cues such as where the horizon is. Such "whiteout" conditions are very similar to the desert phenomenon known as "brownout" experienced by helicopter pilots landing and taking off in terrain where loose sand can obscure outside vision. Maintenance of the aircraft, if required, can also be a monumental task in situations where subzero temperatures are common. When possible, cargo/passenger loads and unloads are done with the engines running, to avoid the difficulties in heating up engine oil to an acceptable temperature to allow starting them. With the end of Starlifter operations, the tasking is now being accomplished by the C-17A Globemaster III.

C-141A, 65-0280, taxis at a base in the Republic of South Vietnam in December 1967, with petal doors open. *MACV photo*

C-141A, 66-7954, at Phu Cat AB in 1971, assigned to the 438th Military Airlift Wing, McGuire AFB, in New Jersey. *Norm Taylor*

C-141A, 65-0237, of the 60th MAW at Korat AB, Thailand, in August 1968, with cargo on a pallet and on a K loader after unloading operations. *Mark Aldrich collection*

C-141A, 66-0154, taxis past revetments, with F-101s and C-47s protected somewhat against Viet Cong air base attacks. *Robert F. Dorr collection*

In the worst accident of the Vietnam War, an A-6 Intruder collided with C-141A, 65-9407. Its load of acetylene cylinders destroyed the plane. *USAF*

Ex-POWs board C-141A, 65-9398, during the process of rehabilitation after being released. Red cross on tail fin signifies a humanitarian mission. *USAF*

Scene of joy and relief as POWs cheer in a C-141 after being released from captivity. Seats face aft when used aboard military planes. *Department of Defense*

The most famous C-141 of all, the "Hanoi Taxi," 66-0177, seen here at Offutt AFB, Nebraska, in 1973. It was the first plane in to pick up former POWs. *Jerry Geer via Bob Esposito collection*

C-141C, 66-0177, at the MacDill AFB (Florida) air show in April 2005. The plane always drew large crowds when it was on static display at open-house events. A POW-MIA flag is placed in the open cockpit escape hatch, and a dedicated electric power cart is connected to the external power receptacle. *Author*

C-141C, 66-0177, "Hanoi Taxi," main instrument panel, showing 6-by-8-inch active matrix liquid crystal displays, part of the -C variant upgrade.
All photos this page, author

A nose angle view of 66-0177 at its current assignment, as a permanent display at the Air Force Museum. It was the last C-141 to fly when delivered to Dayton.

The navigator panel on 66-0177. Various radios are also located here, with GPS being the primary method of plotting courses and waypoints.

The airplane is a major draw to the Dayton facility, and well worth visiting in a gallery full of Air Force history and artifacts of the past.

The left photo shows 66-0177 parked with cargo doors open, as a welcome-home sign is positioned during Operation Homecoming. At right are some of the signatures on the forward cargo cabin wall and the inside of the open lavatory door of those who flew, worked on, and rode as passengers in the "Hanoi Taxi," as well as on C-141s, during its service career. *Left image, USAF; right image, author*

This elevated view of the gallery at the National Museum of the United States Air Force shows C-141C, 66-0177, surrounded by significant aircraft such as past presidential transports in the background, a C-119J Flying Boxcar (*at right*), and an XC-142 behind the left-wing engines. A TF33-P-7 engine is displayed beside the left forward fuselage. Above the "Hanoi Taxi" is a Boeing BQM-94A Gull remotely piloted vehicle. *Author*

Whether in the dense heat and humidity of Southeast Asia or the frozen wastes of the Antarctic, C-141s delivered their cargoes of material and passengers, as this Starlifter of the 7th MAS shows as it unloads at McMurdo Station in support of Operation Deep Freeze. A tracked firefighting vehicle has taken up position about 45 degrees to the nose and maintains watch as the operation proceeds.
Jose Lopez / USAF

C-141B, 65-0259, loads up in Christchurch, New Zealand, for a Deep Freeze support flight as pallets move from a K loader into the cargo hold. *Bob Fehringer / USAF*

C-141B, 65-0239, at McMurdo Station, with a team from the National Science Foundation in red parkas. They are worn to stand out in the white landscape. *Jerry Morrison Jr. / USAF*

A UH-1N is loaded to be used in supporting scientists at the Antarctic base. It has twin engines and a contrasting paint scheme for easy visual sighting. *Joe Cupido / USAF*

Unloading operations are underway as a C-141C has flown in much-needed supplies to the remote outpost during Deep Freeze 2001. *Kim Allain / USAF*

The number 3 engine needs work: a parachute is wrapped around the nacelle. A hose will pump in warm air from the cart in background.
Bob Fehringer / USAF

An Emperor Penguin chick is loaded onto a C-141 for a flight to San Diego to be studied by the National Science Foundation. Ice packs help keep it cool.
Jose Lopez / USAF

C-141B, 66-0183, is unloaded by a forklift on the Pegasus Runway during ongoing support operations in Deep Freeze 1997. *USAF*

A 20,150-pound liquid helium tank is offloaded, to be used in the study of cosmic background radiation by the National Science Foundation.
Jose Lopez / USAF

C-141C, 67-0015, on Ross Ice Shelf runway with brake and engine heaters at work while the cargo load is in the process of being removed.
Kim Allain / USAF

Dick Morita of the Washington Polar Science Center checks one of ten weather buoys, before being air-dropped over the South Pole.
Garfield F. Jones / USAF

C-141s at rest in October 1980, as passengers and cargo are offloaded and equipment and people are taken aboard during the Deep Freeze 1980 support effort. *Bob Fehringer / USAF*

There are times when landings are not necessary. An airdrop of cargo is undertaken over McMurdo Station in June 1983. *Garfield F. Jones / USAF*

CHAPTER 4

C-141s at Work and C-141B Modifications

C-141s proved their worth and reliability on a recurring basis all over the world in many mission profiles, from supporting American policy decisions, to responding in a timely manner to domestic and international calls for aid, to relief flights to provide support in the wake of natural disasters, to military exercise operations with allied nations, to supporting NASA and government officials up to and including the president.

Throughout all of this, it was noted that a C-141 would oftentimes fill up to capacity before the maximum weight-carrying limit was reached. Studies concluded that by stretching the fuselage, plus adding in-flight refueling capability, MAC would gain the increase in airlift capacity equivalent to that of ninety new-build planes. After the first modified plane flew, Lockheed was awarded a contract in May 1978 and was paid $407 million to modify 270 C-141A models, creating C-141Bs in the process. The fuselage would be split open at two locations, with one separation made ahead of the wing, and another aft of the wing box assembly. Each gap would be filled with an addition to the fuselage length, with the forward section or "plug" adding 13 feet, 4 inches of interior cargo space growth. The aft plug would add 10 feet more. By doing this, up to thirteen 463L pallets could be carried, as opposed to ten on the C-141A version. Nearly one-third more internal volume was created, adding 240 square feet / 2,171 cubic feet (62.03 cubic meters). Fuselage length overall increased from 145 feet to 168 feet, 4 inches. Design payload weight grew to 74,233 pounds, and maximum takeoff gross weight increased to 323,000 pounds. Design lifetime flight hours also were increased from 30,000 flight hours to 45,000. The entire C-141B upgrade program cost $475 million.

In three-tone European One camouflage, a C-141B of the 437th MAW takes on fuel from a KC-135R Stratotanker of the 19th Aerial Refueling Wing (ARW). Both planes fly at the same airspeed, and the receiver maneuvers into range while powering down equipment emitting radio frequency energy, such as the weather radar, since it might interfere with the fueling procedure. With the ability to avoid fuel stops, the Starlifter was capable of nearly unlimited range. *Hans H. Deffner / USAF*

SSgt. Ernesto X. Perez briefs Army troops on safety issues before a C-141 loading operation begins at Biggs Army Airfield (AAF), Texas, in June 1981. *Bill Thompson / USAF*

A K loader is parked as a 463L palletized load is pushed into a C-141 of the 8th Airlift Squadron, McChord AFB, for a flight in the course of Deep Freeze 1998. *Jerry Morrison Jr. / USAF*

During Reforger drill 10-82, a K loader is guided in to align with a C-141 ramp so a 463L pallet can be moved aboard and secured. *Bill Thompson / USAF*

With the K loader lined up, a pallet of generators on their way to the South Pole are about to move into a C-141 via roller conveyor channels. *Kim Allain / USAF*

A TAC Loader, or K loader, is lined up with the lowered ramp of a C-141B, as a load configured for airdrop is being pushed aboard by means of roller conveyor channels, seen under the plywood base. Also visible is a layer of cardboard cushion, intended to absorb the impact load once the cargo is released from the aircraft as it flies over a drop zone (DZ) during exercise Big Drop II in May 1995. The K loader in standard configuration can carry up to three 463L pallets. The loading-platform height is adjustable from 40 to 75 inches above the ground, with a maximum lifting capacity of 25,000 pounds. The loader can be modified with extensions to increase pallet loads to five. In so doing, length is increased from 24 feet, 5 inches to 37 feet, 8 inches overall. *David Wilcoxson / USA*

A loaded 463L pallet displayed inside C-141B 64-0626 at the AMC Museum, Dover AFB. The name comes from the date (April 1963) when such pallets were invented. *L* stands for logistics, and the military designation is HCU-6E. Dimensions are 2¼ inches high by 88 inches long by 108 inches wide. They weigh 290 pounds each and have a balsa wood core surrounded by corrosion-resistant aluminum. There are twenty-two tie-down rings for attaching cargo netting to hold loads in place. This pallet is on the closed cargo ramp section of the aft cabin, with the pressure door—closed behind it—that maintains pressurization in the cabin section of the aircraft while it is in flight, folding upward when on the ground. *Author*

Once pallets have been locked in place, inventories of each, with a check of cargo straps, are performed prior to takeoff. The netting used on each pallet is composed of three patterns: two side nets and a top net. Side nets attach to the pallet, while the upper sides connect to the top net. Adjustment points allow netting to be tightened for the duration of flight activity and ground handling. Up to 10,000 pounds of cargo can be secured, up to a height of 96 inches. Pallet side rails on the cargo bay floor prevent lateral or vertical movement up to external forces of 8 g. Photo at right shows a K loader in Germany, as a load of pallets are removed from C-141B, 65-0261. AFRC on the tail fin indicates that this aircraft is assigned to Air Force Reserve Command. *Both photos, Lockheed via Paul Minert*

Diego Garcia in the Indian Ocean. In order for vehicles to drive up the ramp into the cargo cabin, aft ramps are placed in preparation beforehand. *David N. Craft / USAF*

An M-270 MLRS vehicle is driven up inside a C-141B. Plywood sheets prevent tracked vehicles from damaging the ramps and cabin floor as they drive in. *Jesus Villalobos / USAF*

First-time loading of the M-270 multiple launch rocket system of B Battery, 3rd Battalion, 27th Field Artillery, XVIII Airborne Corps. *H. H. Deffner / USAF*

An M-998 HMMWV drives off a C-141B at MCAS Cherry Point, North Carolina. A marshaler provides driver with guidance since his vision is restricted. *John K. McDowell / USAF*

The left picture shows a high-mobility multipurpose wheeled vehicle (HMMWV) being loaded for eventual airdrop. Lashed and chained to a pallet, with parachutes attached, it will join up with airborne assault troops in an exercise scenario. At right is one of the fleet of presidential limousines about to be unloaded. C-141s were routinely tasked with VIP vehicle airlift, many times accompanying the president to summit meetings, state visits, or campaign stops in the continental United States. The vehicles would precede the official delegations, part of the workup to an actual event. *Left photo, Lockheed; right photo, Paul Minert collection*

A maintenance stand is unloaded. To avoid losing control while exiting, a safety strap regulates its momentum until on the tarmac. *Scott Stewart / USAF*

An Army OH-58 scout helicopter, rotors removed and windows taped with plastic, is ready to fly from Fort Lewis, Washington. *Gene D. Tackett / US Army*

M-16 assault rifle rack is loaded through the normal crew entry door by members of the 305th Security Forces Squadron on a flight to Savannah, Georgia. *Kenn Mann / USAF*

An AH-1 Cobra attack helicopter is loaded during an emergency deployment readiness exercise (EDRE). The cargo hoist pulls the chopper into the cabin. *Gene D. Tackett / US Army*

A UH-60A Blackhawk helicopter is in airlift configuration, with rotor blades folded and tail fin folded to the right side. The Army's 101st Airborne Division at Fort Campbell, Kentucky, was testing out equipment use of an air transportability kit in July 1979. Space for a second helicopter allows for two to be moved simultaneously. This is yet another instance of how versatile the C-141 is at handling loads of various shapes and sizes. *W. L. Bonfiglio / USAF*

After Exercise Kindle Liberty '83, M-38 light vehicles are loaded up for the flight from Panama to the United States. They are driven in and chained down. *R. Bandy / USAF*

SSgt. Andrew L. Cobb of the 437th Aerial Port Squadron chains down an ambulance used in the Bright Star '87 event in Egypt. *Scott Stewart / USAF*

Operation Haylift. A C-141B is shown loaded with hay for livestock, to be delivered to the southeastern United States, suffering from an extended drought. *Robert C. Marshall / USAF*

TSgt. Jose Chaudez, loadmaster with the 730th AS, does weight and balance checks with cargo secured. Mission is an airdrop over Fort Benning, Georgia. *Bill Kimble / USAF*

Soldiers from A Company, 2nd Platoon, 5th Recon Battalion, 3rd Marine Division, hook in static lines from their MC-1-1 chutes in November 1999. *Aaron Landegent / USMC*

A combat controller of the 22nd Special Tactics Squadron scans the drop zone from the opened aft ramp during the US/Egyptian exercise Iron Cobra. *Sean Worrell / USAF*

En route to an amphibious airdrop mission, a team of combat controllers get ready checking gear and each other as the drop zone approaches. *Joseph F. Smith Jr. / USAF*

Over DZ Sicily, a paratrooper of the 82nd Airborne Division looks out from the right aft troop door at the scene below as 5,000 soldiers jump at Fort Bragg. *Kenn Mann / USAF*

Before a paratroop drop occurs, air deflectors are installed on the C-141. These reduce the blast of the slipstream airflow around the aircraft as the soldiers depart the plane. In the left picture, Senior Airman Thomas Bacztub (*at left*), works with Senior Airman Joe Sheehan, both members of the 62nd Aircraft Generation Squadron, McChord AFB, to make sure that one is ready for use during joint task force exercise 98-1. Right photo shows a pair of robust actuating arms and electric motor used to operate the device. *Left photo, Bill Kimble / USAF; right photo, author*

A jumpmaster looks at the terrain below in the left picture, part of a prejump safety check over Fort Benning, Georgia, in May 2004, from an open C-141B right aft troop door. He is in charge of his group and ensures that all is in readiness before committing to the jump operation. The right photo shows how covert signals may be transmitted to aircraft delivering assault troops as a signal mirror redirects sunlight to relay a message. Using this method instead of radio traffic, which may alert hostile forces, is combat controller team (CCT) member TSgt. Richard Heins, of the 317th Tactical Air Wing, who has arrived at the drop zone ahead of the main assault force. *Left photo, Bill Kimble / USAF; right photo, Bob Wickley / USAF*

Airdrop scene over the designated drop zone as Starlifters deliver their loads of paratroops. Parachutes are attached to static lines that open them once the soldier clears the aircraft. In addition to troops, equipment such as small vehicles, ammunition, food and water, artillery pieces, and trucks all can be dropped with them to enhance mobility and firepower capabilities. Future airdrops can be called in, or else, once an airfield has been captured or constructed, the aircraft could land and their cargoes unloaded on the ground. *USAF*

Guiding more C-141s in during a mass airdrop operation during NATO exercise Reforger '80 (return of forces to Germany) is an Air Force combat control team member. CCT teams are the "advance party" that makes sure designated DZs are suitable for the drops of troops and equipment, and that hostile resistance is either not a factor or, if it is, can be dealt with before the drop occurs. Smoke grenades are a visual source for where the drop zone is and for the wind direction. The right picture shows a palletized vehicle slowing as it descends under large parachutes. These are dropped close enough to the troops so that they can swiftly secure them and begin the process of unpackaging for use in support of planned objectives. *Left photo, Bob Wickley / USAF; right photo, Lockheed*

A 463L pallet hangs in midair after being rolled aft inside a C-141B cargo compartment. The drogue chute helps pull the load free. Once the load begins a vertical descent, the main chutes open to slow the cargo as it falls gently to Earth. The aircrew makes trim adjustments as these pallets depart the aircraft, making it lighter and changing the center of gravity. The loadmaster advises the pilots over the intercom system as to events unfolding. Timing and spacing of aircraft are vital to ensure that the planning beforehand ends in favorable results to the paratroopers on the ground. *Dean Wagner / USAF*

After an airdrop event in the former Federal Republic of Germany in September 1982, civilians look on as US Army troops have unpackaged a 105 mm howitzer. After removing the netting, parachutes/risers, and cardboard cushions, they can drive off the pallet to their firing location. Airdrops such as these enhance mobility and speed of an assault, confusing the enemy, and, in so doing, exploit the situation to the favor of the airborne forces. *Ken Hackman / USAF*

Over the Republic of Egypt in 1982, a Bright Star scene involves C-141s disgorging paratroops to achieve goals in a desert environment. *USAF*

Wadi D2 in Egypt is the scene as a small team performs a high-altitude–low-opening (HALO) jump profile off the aft ramp during exercise Iron Cobra. *Paul Caron / USAF*

Air Transport Specialist Senior Airman Ricky Gregg plots GPS location as the first C-17A airdrop with C-141s is set for a clearly marked drop zone. *Lisa M. Carpenter / USAF*

Ice Station Zebra, 60 miles from the South Pole, gets this containerized-delivery-system (CDS) string of packages from a passing C-141B. *Jesse M. Villalobos / USAF*

A C-141B drops fuel and supplies to an outpost close to the North Pole. One load is free of the aircraft as another makes its way aft.
Lou Hernandez / USAF

Gallant Eagle '84 included this airdrop of 82nd Airborne soldiers over San Luis Obispo, California, as C-141B, 64-0646, contributes its load of troops.
Izar / US Army

Heavy pallets make their way earthward in exercise Cabanas '86 in Morocon Country, Honduras. Joint drills promote friendship and stability.
Lou Hernandez / USAF

Another Reforger scene as C-141s in 1980 conduct a mass airdrop in the Lhuende DZ in the former West Germany. *Paul J. Harrington / USAF*

C-141s abandoned the white/light-gray gloss paint, taking on this three-tone flat paint camouflage pattern in order to better blend in with foliage when at low level, if viewed from above. This "European One" livery consists of Gray (FS 36118), Green (FS 34102), and Dark Green (FS 34092). The engine pylons and nacelles are of the Gray shade, the radome being of the Dark Green color entirely. The flat colors and black markings resulted from studies such as Project Musketeer, which discovered that gloss paint and flashy squadron insignias on tail surfaces reflect light and heat in sufficient quantities to enhance detection by infrared guided missiles. Aircraft has petal doors open, ready to drop containerized cargo during Volant Rodeo '83. *Ken Hammond / USAF*

Aircrews must also be able to operate in an environment where chemical weapons have been used. A copilot makes takeoff checks in mission-oriented protective-posture (MOPP) gear. *William B. Belcher / USAF*

The MOPP outfit changes with the threat faced. A technician has changed out an attitude indicator in his ensemble. *Ken Hammond / USAF*

In MOPP level 4 gear, SSgt. Ryan Abraham, 731st Air Mobility Squadron, unwraps pallets at Osan AB, Korea, during an exercise. *Chenzira Mallory / USAF*

A C-141B flight engineer completes a checklist while adorned in a chem-warfare suit. Hood, gas mask, gloves, and suit all are required when threat is highest. *Ken Hammond / USAF*

C-141 Timeline, 1968–1984

April 1969: First airlift of a Minuteman III ICBM occurs, from Hill AFB, Utah, to Minot AFB, South Dakota.

May: Two C-141s airlift 50 tons of insecticide to Ecuador in operation Combat Mosquito, in response to an epidemic of encephalitis. Once in-country, the areas were sprayed by UC-123 Providers.

July: C-141s pull 25,000 troops out of Vietnam, airlifting them to McChord AFB.

November–December: After a tropical cyclone devastated East Pakistan, C-141s and C-130s deliver aid amounting to 140 tons, with some flights of 10,000 miles in distance flown.

June–July 1970: C-141s contribute to an airlift of some 23,000 refugees who had fled East Pakistan (Operation Bonny Jack) from Tripura Province in India to Gauhati Province. Aid contributed amounted to 2,000 tons of relief supplies.

February 1972: C-141A, 66-0141, of the 62nd MAW, lands in Peking, in the People's Republic of China, in advance of President Nixon's visit.

February 1973: C-141A, 66-0177, lands at Gia Lam Airport in Hanoi, North Vietnam, to bring home the first planeload of former American prisoners of war in Operation Homecoming. This plane was named the "Hanoi Taxi" in honor of that event.

October–November: C-141s and aircrews for twenty-three aircraft fly in excess of 3,000 hours during a mass airlift event known as Operation Nickel Grass, the response to the attacks by Egypt and Syria on the state of Israel, also known as the Yom Kippur War. In thirty-two days, C-141s as well as other airlifters bring in 22,325 tons of tanks, artillery, and ammunition. Initially, the Israeli national airline El Al begins airlifting duties, but a major shortfall soon becomes obvious. Attempts to have US national civil carriers help out do not bear fruit. On October 12, President Nixon orders the Air Force to use all available assets to deliver war materials. The first loads are en route nine hours later. Using Lajes Field on the Azores, a rest-and-refueling stop is established. KC-135 tankers are also refueling thirty-six A-4 Skywawks and forty F-4 Phantoms eventually flown directly to the Israeli air force to replace attrition losses. Also included are twelve C-130 Hercules transports. Flight paths had to be precisely flown directly over the Strait of Gibraltar, then east down the center of the Mediterranean Sea, avoiding hostile Arab nations to the south, and European allies unwilling to get involved to the north. US 6th Fleet fighters escort all of the aircraft, with the Israelis taking over escort duties once aircraft come within 150 miles of the coastline. Once unloaded, supplies are usually at the zones of conflict within six hours. In the wake of Nickel Grass, the OPEC cartel embargoes oil shipments to America, but also seen is a lack of support bases to streamline the airlift process. This results in more emphasis on aerial refueling. One outgrowth of that strategy is installing aerial-refueling hardware into the C-141B variant, as the C-141A lack of such gear is a point of note in post–Nickel Grass critiques.

April 1975: Over 200 C-141 flights in addition to C-130s remove 45,000 people from Saigon, South Vietnam, with 5,600 of the total being US citizens.

April–September: In a total of 251 C-141 and C-130 flights, plus 349 commercial flights, 120,000 refugees of the Vietnam Conflict are resettled from Pacific island bases to the United States. The first C-141 planeload of Operation Babylift passengers delivers sixty-five children to McChord AFB to be settled with adopted families.

February–June 1976: Guatemala suffers a serious earthquake, to which 927 tons of aid are delivered by C-141s, C-5s, and C-130s. Medical, engineering, and communications workers totaling 696 people are also deployed during Operation Earthquake.

May–June: Guam suffers a strike from a typhoon. In response, C-141s and other transports move aid totaling 2,652 tons to relieve island inhabitants.

January 1977: The first modified C-141A (66-0186), with a lengthened fuselage and aerial-refueling YC-141B variant, is assembled and displayed at the Lockheed Georgia facility.

January–February: A huge snowstorm inundates the Buffalo/Pittsburgh region with accumulations of over 100 inches. Transport of snow removal equipment totaling 1,160 tons and evacuation of 430 people are conducted by all types of airlift models in Operation Snow Blow.

March: First flight of 66-0186, the first C-141B Starlifter. Two 747 Jumbo Jets collide at Los Rodeos Airport, at Tenerife in the Canary Islands. C-141s are sent in to evacuate victims of the disaster to medical facilities in the continental United States.

February 1978: Another major snowstorm buries southern New England. Operation Snow Blow II moves in 2,339 tons of snow removal equipment, generators, and communications hardware while evacuating 1,000 individuals.

May–June: After European nationals are threatened by rebel invaders crossing into Zaire from Angola, Operation Zaire I begins using forty-three C-141s and C-5s to airlift French and Belgian troops with 931 tons of equipment to protect their people. Operation Zaire II brings them out, including 1,619 tons of cargo and 1,225 passengers.

November: C-141s fly to Georgetown, Guyana, to recover 911 victims and five survivors of the Jonestown massacre.

December: In eleven C-141 missions with C-5s, 900 foreign nationals are evacuated from Tehran, Iran, as instability increases there. Flights out continue until February 1979, by which time 5,700 people had been flown out.

March 1979: The Three Mile Island atomic power plant accident occurs near Harrisburg, Pennsylvania. MAC aircraft of all types bring in radiation-monitoring gear, lead shielding, and chemical agents.

August: Hurricanes David and Frederic cause damage in the Caribbean, and relief flights deliver more than 2,000 tons of humanitarian aid.

October: A fire in a Marines enlisted men's barracks in Japan results in two C-141s outfitted for aeromedical evacuation responding to pick up thirty-eight victims, airlifted back to the States for follow-on treatment.

April 1980: A C-141B, in a demonstration of range capability, refuels one time on a flight of 11 hours, 12 minutes, from Beale AFB in California to RAF Mildenhall in the United Kingdom.

May: The 62nd MAW takes its first C-141B, 63-8082.

June 1982: The last C-141B conversion is completed, on the twenty-ninth.

September 1983: In a large operation, called Rubber Wall, C-141s take part in an airlift mission to bring in over 4,000 tons of equipment from the continental US (CONUS) to Marines stationed in Lebanon.

October–November: The United States invades the Caribbean island of Grenada. Operation Urgent Fury is ordered to rescue American students at a medical university, restore democratic rule after a coup, and eliminate a Soviet/Cuban military facility where a large runway had been constructed. C-141s are part of the 496 missions flown, bringing in 7,709 tons of cargo and moving 11,381 passengers.

February 1984: C-141s fly to Larnaca, Cyprus, to bring home Marines who had been on duty in Lebanon.

March–April: C-141s fly seventeen missions as a show of force to support Egypt and Sudan from threats of military moves by Libya.

C-141A, 66-0186, in the beginning stages of conversion into the YC-141B. Two sections of new fuselage plugs are in foreground. *All images this page, Lockheed*

The aft fuselage has been moved away from the wing. Round hole (*upper left*) and missing hatch below it are used for emergency escape.

Removing the outer skin reveals ribs and longerons of the inner fuselage. Work platforms allow technicians to complete work before sections can come apart.

Wing section stands alone at this phase, with both the forward and aft fuselage sections removed. The plugs can now be inserted.

YC-141B, 66-0186, outside after the modifications. Unpainted fuselage sections are the new inserts, with the hump above the cockpit being the fuel receiver housing.

Left side profile of 66-0186 as the YC-141B. With the longer fuselage, the plane could now be filled up to capacity, as it was initially designed for.

The receptacle for taking on fuel in flight is known as the universal aerial-refueling receptacle slipway installation (UARRSI). *All photos this page, Lockheed*

The YC-141B under tow. This plane was the 212th C-141A built, indicated by its assigned MSN of 6212. Wing walkers make sure that areas around plane are clear as it is moved.

Two photos of the YC-141B in flight for the first time on March 24, 1977. Follow-on flight tests would involve airflow studies, in-flight refueling, and cargo handling/drops. *Both photos, Lockheed*

The YC-141B Starlifter, 66-0186, in flight. The *Y* in the designation signifies aircraft developed beyond the experimental stage, but not ready to be classed as a production item. The white upper surfaces resulted in less heat retention, reflecting sunlight. This would change once the planes were painted in the European One flat, three-tone camouflage. This would change once again when the flat-gray overall paint FS 16173 was adopted. *Lockheed*

The YC-141B poses beside C-141A, 65-0269. The -B variant is seen visually as longer, with the midair-refueling housing above "US Air Force." The flat-gray painted section above the cockpit windows is to eliminate glinting and reflections, which can cause a visual challenge to the tanker boom operator as the flying boom is positioned to fit into the receptacle. The Starlifter was on its way to a new chapter in its service career with these enhancements. *Lockheed*

As a T-38 Talon provides a photo platform and two extra sets of eyes to clear the area, the YC-141B moves up to begin a refueling operation while in flight as the KC-135 flying boom is ready to make contact (in the left photo). The right picture is a close-up of the UARRSI housing, with tufts attached. They provide a visual indication of airflow patterns, which are studied to ensure the best-possible aerodynamics in this area of the modification. The receptacle door is open, and the probe can be extended/retracted as it is flown into position. Fuel transfer rate is 23,592 US gallons in twenty-six minutes. *Both photos, USAF*

Col. Watts, 63rd Military Airlift Wing commander at Norton AFB, and the base Organizational Maintenance Squadron commander hold a symbolic key as the first C-141B Starlifter is accepted in a ceremony on July 18, 1980. The aircraft in the background awaits its first tasking as a C-141B version. *USAF*

CHAPTER 5

C-141 Versatility, JACC/CP, KAO, C-141B SOLL II, and C-141C

C-141s never failed to impress, resulting from numerous modifications carried out by the Air Force over many years, from conducting operations in the thick, congested airspaces of low-level nap-of-the-Earth flights, to supporting special-operations teams in darkness, to assisting NASA in loading and installing a unique telescope used at high altitudes in as steady a flight path as possible, to studying regions of outer space. The Starlifter was further modified from the -B variant into the -C model to stay up to date with improved navigation hardware and electronic flight instrumentation systems (EFIS or "glass") displays. Self-defense systems were also incorporated, since the large, slow-moving cargo planes over time became prime targets for attacks primarily from ground-launched, shoulder-fired, surface-to-air missiles. One plane, the first one built, even gave a sterling performance towing a fighter jet, in a test of concepts to launch spacecraft into low-Earth orbit.

JACC/CP: Four C-141Bs (64-0623, 65-0221, 67-0019, and 67-0020) were modified, with the interior cabin altered to house the joint airborne communication center / command post equipment. Extra antennas for expanded HF/VHF/FM and UHF were added to the exterior fuselage, in addition to the standard comms fit. The most obvious visual recognition clue was the two aft-oriented HF antennas, one attached to each wing at about the half-span location on the trailing-edge sections, facing aft.

Kuiper Airborne Observatory (KAO): The single Lockheed L300 commercial example built, MSN 6110/N4141A, having never been ordered into mass production, was eventually given to NASA. The N number was changed to N714NA, with call sign NASA 714. The plane was modified to load aboard and operate the Kuiper 36-inch / 91.5 cm infrared (IR) telescope, a conventional Cassegrain reflector design. The aircraft would fly at altitudes of 41,000 to 45,000 feet, above atmospheric water vapor, which absorbs IR wavelengths. In use, the KAO platform discovered the rings around the planet Uranus in 1977 and also, in 1988, detected the presence of an atmosphere on Pluto. It also observed supernova SN1987A and was able to detect and track the heavy elements of cobalt, nickel, and iron residues.

SOLL II: The modification program began as the early 1990s Project Pathfinder. Planes were adapted to be used in daylight/VFR conditions only, in support of special-operations forces. The SOLL I (special operations, low level) label was eventually applied to them. The follow-on SOLL II variant was modified for night/IFR use, and the thirteen planes were converted by adding the SOFI-Mod (special-operations forces, improvement-modification), at a cost of $41 million. Nearly 1,500 pounds / 680 kg of extra hardware was installed. Included was an AAQ-17 forward-looking infrared (FLIR) sensor turret, located at the chin section of the nose. An image monitor and joystick to operate the turret were placed at the navigator station. Flanking the FLIR housing were a pair of what look like antenna radomes but are in fact aerodynamic fairings, used to house a pair of ALE-40 flare dispenser modules in each fairing. Radar-warning receiver (RWR) installation was of the ALR-69 system. Another warning system used to detect hot exhaust plumes from missiles approaching the aircraft from below is the AAR-44 suite of IR-detecting sensors. They could automatically trigger the release of flares to deceive heat-seeking missile types fired at it. The aircrew had the option of manually firing flares if required. The aircrews themselves had to be airdrop qualified before being considered for SOLL duties. During ground alert standby, they were able to get airborne within an hour of the initial notification. Aircrews were composed of three pilots, one navigator, two flight engineers, and one loadmaster, optimized for the air-land mission profiles, known as J-1. Crew J-2 was made up of three pilots, two navigators, two flight engineers, and two loadmasters for more-involved flights. Crew J-3 was the same as crew J-2 but was dedicated to missions involving boat-drop objectives.

Flights were mostly carried out at night, at low level, and if on a training mission, extra aircrew members would be aboard to maintain proficiency in phases of flight such as dry or wet hookups to an aerial-refueling tanker, for example. During low-level flight, the aircrews wore AVS-9 night vision goggles (NVGs). No exterior lighting was on, since the aircraft flew low and fast on an ingress course to a designated drop zone in denied territory, where anything from ammunition, armament, and food/water to even light vehicles could be dropped in covert operations to support special-forces teams on the ground. Once an airdrop was concluded, the aircraft executed a preplanned egress route, avoiding any known threats while aircrews and loadmasters scanned the airspace and ground below for any hostile activity. Approaches and landings were also performed with lights out and wearing NVGs. The SOLL II planes operated by the 16th Airlift Squadron bolstered special-ops airlift capabilities of MC-130E and MC-130H Combat Talon Hercules assets. The SOLL II modification was a radical departure from normal Starlifter routines, but one at which the C-141 excelled on numerous occasions.

C-141C variant: As the Starlifter fleet aged, new technology in air traffic control coordination and identification, plus GPS system enhancements and glass display use, resulted in an upgrade applied to sixty-three of the lowest-time airframes. New component items were also included, replacing dedicated mechanical as well as electromechanical devices. An all-weather flight control system (AWFCS) with digital autopilot eased aircrew workload. Improved satellite communications (satcom) hardware was fitted, and a ground collision avoidance system also was incorporated into the -C model enhancement package. The last flight of any C-141 was of a -C conversion, 66-0177, flying to Dayton, Ohio, on May 6, 2006, to take its rightful place in the National Museum of the United States Air Force.

A C-141, its flight from the United States over, taxis in at the Rhein Main Airbase, West Germany, in November 1984. The marshaler has his wrists crossed, the universal signal for the pilot to apply brakes and stop. At night, light wands are used for enhanced visibility. With the plane parked, chocks will be placed in front of and behind the tires, external power will be connected, and the crew will open doors as necessary if unloading is to take place. *Glenda Pellum / USAF*

TSgt. Terry Netts, *left*, and MSgt. Gregory King, both of the 445th Aircraft Generation Squadron, home-based at Wright-Patterson AFB, Ohio, position a cradle underneath number 3 engine, which needs a change-out in the left picture. This will be raised to mate with the power plant and be secured to it. The engine connections to the pylon will all be disconnected, and the engine can then be lowered. The right photo shows part of a preflight check in April 1998: a visual inspection performed by Senior Airman Joe Glass of each engine intake for cracked or bent blades, foreign objects in the section, and any leaks of fuel or oil. In this scene, engine number 2 is being given the once-over at RAF Mildenhall in the UK. The row of visible blades adjacent to the cone or spinner in the center are fixed in position and do not spin. *Left photo, Ken Bergmann / USAF; right photo, Brad Fallin / USAF*

TSgt. David Weaver does a check of a flap actuator by viewing it and handling the arm itself. Such is the routine of the preflight checklist. *Brad Fallin / USAF*

Nitrogen is used to fill a main gear strut oleo. As the gears extend, the oleo doors open, allowing the units to protrude outside the main gear fairings. *USAF*

Engine oil level and quality are examined by SSgt. Jimmy Olson while on a stop at Incirlik AB in Turkey. Access panel on side of nacelle makes this relatively simple. *Fernando Serna / USAF*

Connecting external power cord is SSgt. Eliezer I. Sosa Jr. of the 815th Air Mobility Squadron at Travis AFB. C-141 electrical systems can now be used without engines or APU. *Richard Kaminski / USAF*

The pilot, Lt. Col. "Hawkeye" Pierce (*left*), with MSgt. Don Boudinet behind him. Lt. Col. Chuck Hanks is copilot. Behind him is the flight engineer, SMSgt. Steve Bell. All are on a Project Trans Am flight in a C-141C model of the Starlifter. Visible are the flat glass panel displays, and the glare shield green screens are display avionics maintenance units. All are from the 89th Airlift Squadron, Wright-Patterson AFB. *Lance Cheung / USAF*

Lt. Col. Samuel F. Oglesby, 702nd Airlift Squadron commander, monitors systems as a slow-ascent approach is made to a KC-10A Extender tanker. *Scott H. Spitzer / USAF*

Cockpit scene as a C-141 flies to Syria to pick up former hostage Joseph Cicippio, held captive in Lebanon over five years. Capt. Gregory Gibbs is the pilot. *Perry J. Heimer / USAF*

First refueling by an AFRC aircrew, as attention is focused by 315th MAW airmen nearing a KC-135E Stratotanker that has flying boom extended. *William W. Magel / USAF*

US vice president George H. W. Bush is in the navigator seat as an observer. Call sign with this guest aboard is Air Force Two. *Donald L. Wetterman / USAF*

Lt. Col. Juan Sotomayor flew nearly 9,000 flight hours in a forty-year career. His last flight ended with the traditional soaking ritual! *Kenn Mann / USAF*

US Navy pilot Lt. Jeffrey Zahn of VF-35 relaxes at the navigator station. His plane was shot down during Desert Storm, and he had been held captive. *Susan Carl / US Navy*

C-141A, 66-0194, at an air show. They proved popular since spectators could go inside them, actually see what they could do, and talk to the aircrews. *Lockheed*

President Ronald Reagan talks to a C-141 aircrew after their efforts in Operation Haylift II while visiting South Carolina. *David McLeod / USAF*

C-141s were used to evacuate American students out of Grenada during Operation Urgent Fury. They line up at Point Salines airfield. *Department of Defense*

An 82nd Airborne soldier escorts students to a waiting C-141B in November 1983 during the final phases of the Urgent Fury invasion of Grenada. *Gary Miller / US Navy*

Students from St. George's University ready to fly out to Charleston AFB, since evacuating the American citizens was a priority. *US Navy*

Troops board a C-141B at Point Salines Field as Operation Urgent Fury winds down. The Starlifter was a valued asset in moving cargo and soldiers during the event. *M. J. Green / USAF*

C-141s have also had the task of bringing home loved ones who lost their lives in the line of duty. The left picture shows victims of the US Army 3rd Battalion, 502nd Infantry, 101st Airborne Division, at rest in a 437th MAW Starlifter after a charter plane crashed in Gander, Newfoundland, in December 1985. The right photo shows pallbearers exiting the cargo cabin after loading the casket of an Army major who lost his life in East Berlin. All the armed services have longtime honored and established procedures steeped in tradition and respect for fallen members, carried out precisely by elite teams that regularly practice their roles in such ceremonies. *Left photo, Vincent R. Kitts / US Army; right photo, Fernando Serna / USAF*

A C-141 arrives in California with US representative Leo Ryan and colleagues killed in the Jonestown, Guyana, incident in January 1978. *USAF*

C-141B, 65-0245, is the backdrop as a joint-services honor guard carries an unknown soldier of the Vietnam War to a waiting hearse. *Dennis G. Plummer / USAF*

The crew remains of STS-51L *Challenger* are brought onto a C-141B at Kennedy Space Center for a flight to Dover AFB in 1986. *NASA*

Firefighters douse hotspots on a C-141B after it was involved in a fire at Pope AFB in May 1993. The view shows what remains of the right wing. *Cindy Burnham / USAF*

C-141B JACC/CP, 64-0623, taxis in after landing at Patrick AFB. At this angle, the only recognition item is the satcom antenna on the top fuselage / wing join fairing. *Author*

C-141B JACC/CP, 66-0221, with nonstandard BuNo marking on tail fin. Satcom antenna is a DM C34 unit for the UHF satellite terminal system (USTS). *Author*

C-141B JACC/CP, 64-0623, left-wing HF probe antenna. The left-wing example shown here is number 1. Right-wing antenna is number 2. *Author*

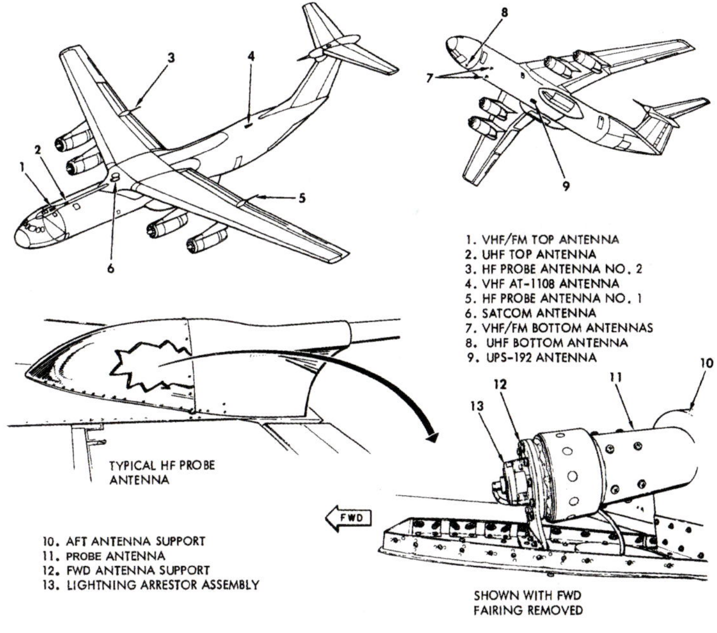

JACC/CP drawing showing antenna locations for HF, UHF, and VHF additions. Bottom two diagrams show the HF mounting in detail. *USAF*

Volant Rodeo events are held annually to test the skills of airlift crews in the spirit of competition, safety, and efficiency. *Rozalyn Dorsey / USAF*

A C-141B drops a pallet during a competition. Location is Pope AAF, in 1992. Accuracy and timing are key to being the best at this phase of the event. *USAF*

TSgt. Dale Wenn leads SrA. Bart Larson, both of the 305th Aerial Port Squadron, as they run from a C-141 during an engine run onload/offload event. *Jeffrey Allen / USAF*

With another Starlifter behind, a palletized load is extracted during a Volant Rodeo airdrop action event. Evaluators on the ground and in the air rated crews on several factors. *USAF*

A C-141 was used to transport an Airstream trailer containing the Apollo 11 crew. The modified unit was known as the mobile quarantine facility (MQF). *NASA*

Four BGM-109 ground-launched cruise missiles, inside CNU-308 containers, are being loaded at Sigonella, Sicily, after Operation Desert Storm. *April S. Hatton / US Navy*

A satellite communications antenna is unloaded from a C-141 in East Africa. It can be assembled and linked to satellites anywhere in the world. *R. S. Mallard / USAF*

AGM-65B scene magnification missiles are offloaded at the conclusion of exercise Sandgroper '82. Munitions such as these are used by tactical aircraft such as A-10s. *USAF*

A K loader in Saudi Arabia unloads a C-141B during the Desert Shield buildup. The first aircraft into the kingdom after the Kuwait invasion was a Starlifter. *USAF*

C-141B, 64-0626, was one of several planes pressed into use during Desert Shield/Storm that were unpainted due to demands on airlift. This plane is now at the AMC Museum. *USAF*

A C-141B of the 164th Air Wing, 155th Airlift Squadron, Tennessee Air National Guard, awaits cargo at Sigonella, Sicily, during Operation Enduring Freedom. It was one of several staging bases for cargo.
Ken Bergmann / USAF

An unpainted Starlifter at a stopover for crew rest at Patrick AFB. The Gulf Wars plus the wars on terrorism accelerated wear on the C-141 fleet. *Author*

Lockheed Model 300, N4141A, used as a company demonstrator before becoming the Gerard P. Kuiper Airborne Observatory. *Lockheed*

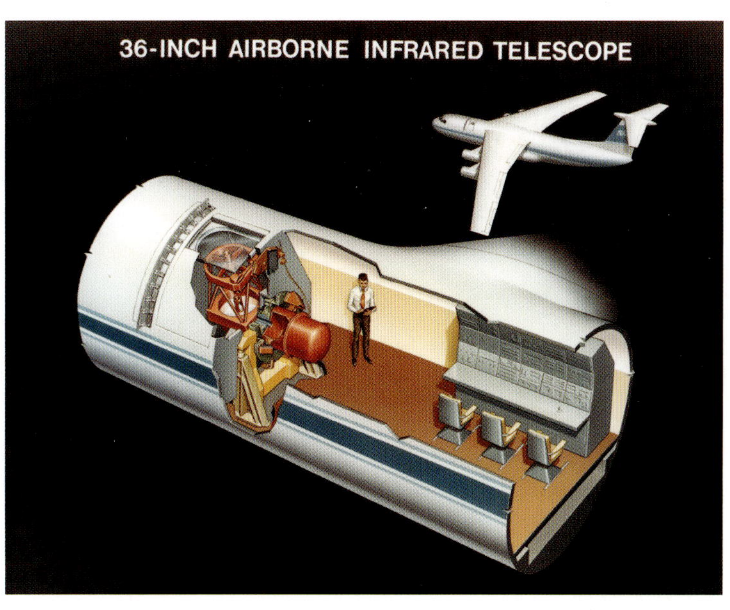

Diagram of the Kuiper observatory interior, with the telescope, optical panel in cabin wall, operator consoles, and technician figure for scale. *NASA*

The Kuiper Airborne Observatory (KAO) at the NASA Ames Research Center. Telescope is behind panel seen partially open. *NASA*

A view of the telescope, with the panel slid open. To the left is a slipstream deflector, used to reduce the effects of turbulent airflow in the opened fuselage cavity. *NASA*

The C-141 underwent testing to investigate wingtip wake vortex creation, and how it could interfere with trailing aircraft and when dropping payloads.

Smoke generator at work under right wingtip. Pipe above partially set flap is a fuel vent line. Lowered wing indicates that plane is in a gentle right turn.

Smoke generators were lit, then the smoke emitted was observed and filmed as the plane flew some prearranged flight profiles. *All photos this page, Paul Minert collection*

View looking aft from the plane as a dropped load descends under a parachute; the wingtip currents, made visible by the smoke trail, show no vortex hazards in this scene.

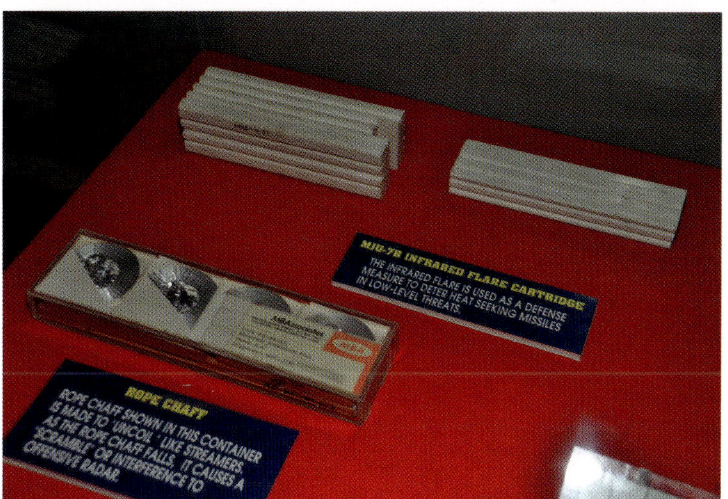

The C-141B type was certified in Project Snowstorm, an outgrowth of a previous project named Pacer Snow, to carry two defensive aids for protection against infrared surface-to-air missiles and IR air-to-air variants fired from hostile aircraft. The upper and lower-left pictures show the AAR-47 missile approach warning sensor (MAWS). Four of these (two forward fuselage-mounted, two on the tail fin) give 360-degree airspace coverage. Each one stares into a sector as the aircraft flies, and when the hot exhaust plume of a missile in cooler air is detected, it alerts the aircrew with an audible warning heard in the headset, and a visual one on a display in the cockpit. The system can then automatically trigger the release of flares, or the aircrew can do this manually. The upper right photo shows the left main gear fairing with two housings for ALE-40 thirty-round flare modules. Two more on the other fairing give a total of 120 MJU-7 flares that can be dispensed in an effort to protect the aircraft from a missile strike. The flares themselves are shown at lower right. They burn extremely hot, and this can cause the missile seeker to lock onto this source as the target, instead of the aircraft itself. Most missile impacts of this type involve hits to the engines, where the jet exhaust is a prime heat source. Their warheads are small, but the damage from them can be extensive, since these explode in a 360-degree arc. This can rupture fuel or hydraulic lines, destroy an engine, or cause flight control problems or loss of cabin pressure. *Upper left, lower left, and lower right photos, author; upper right photo, John Vadas*

In the aftermath of the *Challenger* disaster, C-141B, 63-8085, had its left aft troop door access modified with a facsimile orbiter entry hatch. *NASA via Dennis Jenkins*

C-141B in flight with the pole extended. The hatch forward of this has been removed so photos can be taken. The pole allowed astronauts to egress safely from an orbiter in level flight. *NASA*

If an orbiter could not return to launch site, the astronauts would escape by attaching a ringed loop to the pole. They would then free-fall clear. *NASA via Dennis Jenkins*

The pole in use. Trajectory is intended to go down and aft to avoid striking the left wing and undersides once the orbiter is in a straight and level attitude. *NASA via Dennis Jenkins*

C-141B SOLL (special operations, low level) 64-0630; this adaptation of the Starlifter shows the black AAQ-17 FLIR turret below radome. In many cases the turret was not installed, since there were not enough initially for each plane to have one permanently. Fairing aft of radome contains the ALE-40 flare dispenser modules. Above fairing is one of four ALR-69 radar-warning-receiver aerials. The AAR-44 infrared missile launch detector array and bottom ALR-69 antenna are just aft of nosewheel door on the belly. The "Spirit of TSgt. Paul E. Yonkie" honors the USAF C-141 flight engineer who was wounded by insurgents who targeted the aircraft, C-141A, 65-9408, of the 437th Military Airlift Wing, during an attack on Udorn Royal Thai Air Force Base on July 26, 1968. Sgt. Yonkie was airlifted to Clark AB in the Philippines but passed away as a result of his injuries on September 1. He was buried with full military honors at Arlington National Cemetery in Washington, DC, on September 10. *Photo and historical account by John Vadas*

C-141B SOLL, 67-0014, at Patrick AFB, Florida, in June 2001, has no AAQ-17 FLIR. Aircraft is in the overall flat-gray paint, with external power cord connected. *Author*

C-141B SOLL, 67-0003, with two items of defensive gear shown: the pair of ALE-40 flare modules in the aft gear pod, and, below the troop door, the AAR-44 infrared exhaust plume sensor unit. *Author*

AAR-44 infrared detection array on belly of a C-141B SOLL. Fixed aerials are sensitive to radio frequency (RF) emissions, while the spinning sensor at right detects IR tracks of missile firings. *Author*

Flares erupt from a C-141B SOLL in response to a simulated missile attack. Dispersal pattern is wide, to create a large, hot heat source the missile will home in on. *Gary Ell / USAF*

C-141 Timeline, 1984–2004

May 1984: C-141s deliver 22 tons of medical aid to Peshawar, Pakistan, to support Afghan refugees who had fled war and famine.

July: A C-141 brings thirty-nine passengers of TWA flight 847 from Damascus, Syria, to Rhein-Main AB, who are met by Vice President George H. W. Bush.

December 1985**–January** 1986: C-141s are part of a twenty-six-flight operation to retrieve 248 paratroopers who lost their lives in a DC-8 charter plane crash at Gander, Newfoundland. In addition, they move 770 passengers and 125 tons of cargo.

July: Operation Haylift moves 636 tons of hay (in excess of 19,000 bales) to drought-stricken farmers in the American Southeast. Twenty-four C-141s take part.

July 1987**–November** 1988: C-141s move minesweeping equipment and technicians to the Persian Gulf as Operation Earnest Will escorts reflagged Kuwaiti oil tankers.

April: C-5s and twenty-two C-141s move 1,300 security personnel to Panama to counter political instability, protecting a large American community living there.

December 1989: Women serve on C-141s as aircrew members during missions including airdrop flights.

June 1990: Wing cracks at the station 405 inner/outer wing joints are found, monitored, and repaired as needed.

August 1990**–June** 1991: C-141s fly the most airlifts in Desert Shield/Storm (7,047 out of 15,800), carrying 41,400 passengers and 139,600 tons of cargo and helping move 1,000 tons of fire suppression hardware into Kuwait to put out numerous oil fires set by retreating Iraqi forces.

April 1992: Operation Provide Comfort moves 7,000 tons of humanitarian aid to stranded Kurdish refugees after Operation Desert Storm ends.

May–June: In Operation Sea Angel, C-141s help move 3,000 tons of aid to Bangladesh after a typhoon causes massive damage.

July: Lockheed is awarded $118 million for 121 new wing boxes to replace ones damaged by corrosion from air-conditioning condensation.

September: A flat-gray exterior paint, plus new interior fabrics and panels for interiors, is approved for enhancements over a five-year span.

November: The 100th Afghan relief mission is completed as a C-141 transports ten patients to the US. Two C-141s (65-0225 and 66-0142) collide over north-central Montana while on a training mission with KC-135s. Thirteen aircrew members lose their lives.

October 1993: Operation Restore Hope involves C-141s moving 1,300 troops, forty-four Bradley Fighting Vehicles, and eighteen M-1 Abrams Main Battle Tanks into Somalia after attacks on UN food distribution efforts.

July 1994: Air Wings of twenty-two squadrons airlift supplies into Uganda and Kenya, as Operation Support Hope brings in 3,660 tons of cargo.

June–August 1995: Operation Quicklift moves British and Dutch troops of a UN rapid-reaction force into Croatia.

May 1996: With the award of a contract of $16.2 billion to build eighty more C-17A Globemaster III cargo planes, plans are set for the eventual retirement of the Starlifter fleet.

March 1997: In fifty-seven flights, transporting 532 passengers, C-141s contribute to the airlift of foreign nationals, including Americans threatened in Zaire, in Operation Guardian Retrieval.

December: Typhoon Pak hits Guam. C-141s help bring in over 2.5 million pounds of cargo to aid in relief efforts.

January 1999: C-141s get an upgrade with the installation of all-weather landing system (AWLS) and a digital automatic flight control system, consisting of a display and ground collision avoidance hardware.

February: A C-141B moves Marines to Europe to support NATO air operations in the former Yugoslavia.

September: East Timor declares itself independent, and war breaks out there. A C-141 takes part in the airlift Operation Stabilize to bring in peace-keeping forces from Australia.

August 2000: A C-141 moves 170 troops to Boise, Idaho, to combat eleven fires that had burned 200,000 acres of forest. Four more planes move 600 Marines in to help contain the wildfires.

December: C-141s save relics of the Cold War era, transporting sections of the Berlin Wall and vehicles used in the former Soviet Zone.

January 2001: The C-141C simulator is activated at Wright-Patterson AFB.

April: The last Starlifter departs McChord AFB when it flies to the Davis-Monthan Boneyard in Arizona, ending thirty-six years of C-141 activity there.

September: C-141s bring in medical aid, blood, twenty-two surgical teams, and FEMA staff to McGuire AFB after the September 11 attacks.

November: A C-141 flies in eleven pallets with blankets totaling 18,000 pounds to Ramstein AB, for eventual airdrop to Afghans suffering from exposure to cold temperatures.

January 2002: One active USAF squadron now flies C-141s. Reserve units are flying 100 planes. Air National Guard facilities in Tennessee and Mississippi are equipped with them.

January: The "Hanoi Taxi," 66-0177, returns to flight status after a two-year refurbishment as a C-141C model, assigned to the 445th Airlift Wing at Wright-Patterson, Ohio. A C-141 flies the first twenty Taliban and Al-Qaeda terrorists to Guantánamo Bay, Cuba.

February: C-141s are used to move in the NASA response team to Texas from Florida after the in-flight breakup of space shuttle *Columbia*.

February 2003: A C-141 deploys for two weeks in an Operation Deep Freeze support effort, one of the last in which a Starlifter would be involved.

October 2004: The C-141 formal training unit at Wright-Patterson is closed. C-141 phaseout is scheduled for 2006.

C-141B of 71st Airlift Squadron with a KC-10A Extender of the 79th AREFS takes on fuel over California.
John McDowell / USAF

C-141C Starlifter cockpit, main instrument panel, pictured in September 2000. Upgrade program was similar to the Pacer Crag update to the KC-135 fleet. Glare shield has two pairs of green screens flanking the engine-fire T handles. These are for the display avionics maintenance units. On top of the T-handle section is a defensive-systems AAR-47 MAWS panel. Vertical tape gauges between throttles are the engine performance indicators. In front of the control yokes are a pair 6-by-8-inch active matrix liquid crystal displays (AMLCD). The small lit display to the right of left AMLCDs is a multifunction standby indicator, a backup in case the main displays fail. Below standby indicator, obscured by left-hand throttles, is the Traffic Collision Avoidance System (TCAS) display. At lower center of photo, by aircrew seats, are two flight management system (FMS) interface panels to input data such as waypoints, radio comms frequencies, and fuel burn calculations. *Lance Cheung / USAF*

For Project Eclipse, NC-141A, 61-2775, had its petal doors removed so that the tow cable had plenty of room to account for left/right up/down stresses. *NASA*

QF-106A, 59-0130, rotates under tow. Plane had a shortened nose pitot boom, and tow rope attachment device was placed just forward of the center frame of the canopy windows. *NASA*

The first towed flight was on December 20, 1997. It was eighteen minutes in duration, at an altitude of 10,000 feet. QF-106 engine was at idle throughout towing phases. *NASA*

QF-106A, 59-0130, towing mechanism, side view. It had an electrical disconnect feature as well as a mechanical backup pull cable. *NASA*

Project Eclipse was a joint NASA / Kelly Space Technology (KST) effort to prove that a future aerospace vehicle could be towed to higher altitudes and released. At that time, the vehicle would ignite a rocket engine to climb into a low-Earth orbit. The F-106 was chosen since it had a wing design close to that of the concept vehicle under study. Two Delta Darts, a primary and a backup plane, were acquired from the Boneyard (59-0130 and 59-0010, the latter of which was never flown). The NC-141A was chosen since it needed the least amount of work to modify for the tests. Tow rope was 1,000 feet of Vectran material. For takeoff, the NC-141 would tension the line by moving slowly down the runway. Then it would accelerate and lift off at 120 knots. The QF-106, engine at idle, would rotate at 130 knots and climb at 165 knots. After cable release, the QF-106 would land normally. Six flights were conducted from December 1997 to February 1998. The left picture shows an angle view of the tow cable release device on the QF-106, and at right, an elevated view of the two aircraft connected together. *Both photos, NASA*

The end for many Starlifters came at AMARC, the aerospace maintenance and regeneration center at Davis-Monthan AFB (the Boneyard) in Arizona. The climate there is ideal for long-term storage of aircraft after their service lives have concluded. They are a valued source of spare parts for planes flying, and if no longer needed, they are cut up for salvage and being made into the proverbial "frying pans." The C-141 was one aircraft that performed its role from the beginning as intended. It was further adapted to carry more for longer flight durations, and it did that as well. It was, in fact, in the C-141A variant—shorter in fuselage length than the DC-8 or C-133—that less tonnage was carried than in the 707-320B (34 tons versus 44.9 tons); the C-141A had less range and slower economy cruise speeds than a DC-8 or 707. Outsized cargo that a C-133 or C-124 could haul, the C-141A could not. But what made the Starlifter special was what the plane provided the USAF at the right moment: a jet-powered pure cargo transport with an easily accessible cargo cabin. In the long term, it did carry more outsized cargo payloads than any other plane of that era could hope to carry. C-141s were at the forefront of natural disasters, shows of force, and outright conflicts, carrying tons of war material as well as goods to sustain a combat force or refugees in need of humanitarian aid. It was used in many scenarios, from hauling pallets loaded with anything they could stack on them, to troops and special-operations teams and their unique gear, to airborne paratroopers, vehicles, and artillery pieces. Unlike many aircraft designs, which make it to their full potential only a short time before becoming obsolete, the Starlifter did not follow that path. It worked from the outset and was improved and altered to fulfill other desires. From tests to towing, and from low-level operations to supporting outposts in desolate Antarctic wastelands, to mass movements of cargo to hotspots the world over, the C-141s gave all they had to the war fighters, to those in need, and to the American taxpayers. Although the C-141 is mostly forgotten, since the C-17A Globemaster III has been in the Starlifter's role for some time now, for those who flew, maintained, and were directly affected by the C-141, its place in aviation history and the impact it had will not only be remembered fondly, but be accurately and correctly judged as the right plane at the right time, doing a splendid job in the process. *Dennis Jenkins*